Feminist Cultural Studies of Science and Technology

Feminist Cultural Studies of Science and Technology challenges the assumption that science is simply what scientists do, say or write: it shows the multiple and dispersed makings of science and technology in everyday life and popular culture.

The first major guide and review of the new field of feminist cultural studies of science and technology provides readers with an accessible introduction to its theories and methods. Documenting and analysing the recent explosion of research which has appeared under the rubric of 'cultural studies of science and technology', it examines the distinctive features of the 'cultural turn' in science studies and traces the contribution feminist scholarship has made to this development. Interrogating the theoretical and methodological features, it evaluates the significance of this distinctive body of research in the context of concern about public attitudes to science and contentious debates about public understanding of and engagement with science.

Maureen McNeil investigates three main tropes of the public making of science: making heroes, telling stories and witnessing spectacle. Within this framing she offers a collection of studies of science and technology in contemporary Western culture, including examinations of scientific heroes, new reproductive technologies, military technology and technological tourism.

Transformations: Thinking through Feminism

Edited by:
Maureen McNeil, *Centre for Gender and Women's Studies, Lancaster University*
Lynne Pearce, *Department of English, Lancaster University*

Other books in the series include:

Transformations
Thinking through feminism
Edited by Sarah Ahmed, Jane Kilby, Celia Lury, Maureen McNeil and Beverley Skeggs

Thinking through the Skin
Edited by Sara Ahmed and Jackie Stacey

Strange Encounters
Embodied others in post-coloniality
Sara Ahmed

Feminism and Autobiography
Texts, theories, methods
Edited by Tess Cosslett, Celia Lury and Penny Summerfield

Advertising and Consumer Citizenship
Gender, images and rights
Anne M. Cronin

Mothering the Self
Mothers, saughters, subjects
Stephanie Lawler

When Women Kill
Questions of agency and subjectivity
Belinda Morrissey

Class, Self, Culture
Beverley Skeggs

Haunted Nations
The colonial dimensions of multiculturalisms
Sneja Gunew

The Rhetorics of Feminism
Readings in contemporary cultural theory and the popular press
Lynne Pearce

Women and the Irish Diaspora
Breda Gray

Jacques Lacan and Feminist Epistemology
Kirsten Campbell

Judging the Image
Art, value, law
Alison Young

Sexing the Soldier
Rachel Woodward and Trish Winter

Violent Femmes
Women as spies in popular culture
Rosie White

Feminist Cultural Studies of Science and Technology

Maureen McNeil

LONDON AND NEW YORK

First published 2007
by Routledge
2 Park Square, Milton Park, Abingdon, Oxon OX14 4RN

Simultaneously published in the USA and Canada
by Routledge
270 Madison Avenue, New York, NY 10016

Routledge is an imprint of the Taylor & Francis Group, an informa business

Transferred to Digital Printing 2009

© 2007 Maureen McNeil

Typeset in Baskerville by
Taylor & Francis Books

All rights reserved. No part of this book may be reprinted or reproduced or utilized in any form or by any electronic, mechanical, or other means, now known or hereafter invented, including photocopying and recording, or in any information storage or retrieval system, without permission in writing from the publishers.

British Library Cataloguing in Publication Data
A catalogue record for this book is available from the British Library

Library of Congress Cataloging in Publication Data
McNeil, Maureen.
  Feminist cultural studies of science and technology / Maureen McNeil.
    p. cm.
  1. Feminism and science. 2. Science–Social aspects. 3. Technology–Social aspects. 4. Culture–Study and teaching. I. Title.
  Q130.M36 2007
  306.4'5–dc22   2007025348

ISBN 978-0-415-44537-5 (hbk)
ISBN 978-0-203-93832-4 (ebk)

Contents

Acknowledgements   vii

1   I woke up one morning and discovered that I was doing feminist cultural studies of science and technology   1

2   Feminist cultural studies of science and technology: roots and routes   11

PART I
Making heroes   **25**

3   Newton as national hero   27

4   Making twentieth-century scientific heroes   44

PART II
Telling stories   **69**

5   New reproductive technologies: stories of dreams and broken promises   71

6   Telling tales of reproduction and technoscience   94

PART III
Witnessing spectacle   **111**

7   National and international spectacle: Gulf War I   113

8	Techno-tourism in Florida: American dreams, technology and feminism	132
9	Conclusion	147
	Notes	153
	Bibliography	166
	Index	182

Acknowledgements

This book has been a very long-term undertaking. Hence, it is almost impossible to acknowledge all the contributors to its production. Nevertheless, I do want to take this opportunity to acknowledge the institutional backing and collective environments, as well as the generous encouragement, challenging stimulation and loving support of various individuals that have made this book possible.

I am grateful for various forms of institutional support which have been crucial to this project. The Max Planck Institute for the History of Science in Berlin welcomed me as a visiting scholar from October 2000 to January 2001 and provided the time and space for me to envisage this project in a wonderful setting. An AHRB (Arts and Humanities Research Board) UK research leave grant in 2005 enabled me to assemble the book manuscript into its almost final form. The final redrafting was undertaken while I was a principal investigator on the Media, Culture and Genomics Project in the ESRC CESAGen (Centre for Economic and Social Aspects of Genomics) at Lancaster University.

As well as benefiting from these forms of institutional sponsorship, I had the privilege of working in unique settings and within distinctive collectives in which the ideas for this book incubated. Hence, I am grateful to my colleagues and students in the Cultural Studies Department (originally the Centre for Contemporary Cultural Studies) at the University of Birmingham and in the Institute for Women's Studies, and the Centre for Science Studies at Lancaster University, whose stimulation and challenges have very much influenced my thinking and writing. In addition, this project derives from my engagement with the *Radical Science Journal* collective, with the editorial board of *Science as Culture* and with various feminist activities.

I would particularly like to thank those who have provided feedback on various drafts of chapters and outlines of the book: Wendy Faulkner, Sarah Franklin, Joan Haran, Jenny Kitzinger, Sonia Liff, Nayanika Mookerjee, Michal Nahman, Kate O'Riordan, Celia Roberts, Jackie Stacey and Karen Throsby.

This book derives from endless conversations, discussions and exchanges with colleagues. For these, I thank Sarah Ahmed, Rosemary Betterton, Claudia Castañeda, Andrew Clements, Anne-Marie Fortier, Sarah Franklin, Ann Gray, Michael Green, Joan Haran, Donna Haraway, Richard Johnson, Jenny Kitzinger, John Law, Les Levidow, Gail Lewis, Sonia Liff, Jackie Litt, Celia Lury, Nina Lykke,

Adrian Mackenzie, Nayanika Mookerjee, Maggie Mort, Michal Nahman, Kate O'Riordan, Moira Peelo, Lynne Pearce, Birgit Reinel, Vicky Singleton, Beverley Skeggs, Jackie Stacey, Deborah Steinberg, Fred Stewart, Lucy Suchman, Marja Vehvlainen, Judy Wajcman, Brian Wynne and Bob Young.

My 'women's groups' and the members of my various 'families' have kept me and this book going. So, I offer a special thanks to my 'sisters' for their support and encouragement: Judy Brown, Lesley Caldwell, Ann Gray, Béatrice Dammame-Gilbert, Lorraine Daston, Pat Dyehouse, Jennifer FitzGerald, Christine Hardy, Janet Holland, Susan Lee, Jackie Litt, Mica Nava, Gianna Pomata, Laura Quinn and Janice Winship.

I also thank Adi Kuntsman, who did the final formatting of this book and helped me finally see it off.

Various parts of the chapters of this book have been aired in diverse conference, seminar and lecture presentations that I have given over many years in different locales around the world. While I am unable to list (or even recall all of) these events here, I do want to register my appreciation to these unnamed listeners and unspecified audiences for their comments and feedback.

More specifically, versions or some parts of Chapters 2, 3, 4, 5 and 8 have appeared in earlier publications. I am grateful for permission to reproduce parts of these earlier publications as listed below:

- 'Roots and routes: the making of feminist cultural studies of technoscience', in A. Smelik and N. Lykke (eds) *Bits of Life: Feminist Studies of Media, Biocultures and Technoscience*, Seattle: University of Washington Press, forthcoming, pp. 27–52.
- 'Newton as national hero', in J. Fauvel, R. Flood, M. Shortland and R. Wilson (eds) *Let Newton Be! A New Perspective on His Life and Works*, Oxford: Oxford University Press, 1988, pp. 223–39.
- 'New reproductive technologies: dreams and broken promises', in M. McNeil and S. Franklin (eds) *Science as Culture* (special issue on procreation) 3, pt 4, no. 17, 1993: 483–506.
- 'Telling tales of reproduction', in A. Cervinkova and K. Saldova (eds) *Science Studies Opens the Black Box: Spring School of Science Studies Proceedings*, Prague: Institute of Sociology of the Academy of Sciences in the Czech Republic, 2006, pp. 19–40.
- 'Techno-triumphalism, techno-tourism, American dreams and feminism', in S. Ahmed, J. Kilby, C. Lury, M. McNeil and B. Skeggs (eds) *Transformations: Thinking through Feminism*, London: Routledge, 2000, pp. 221–34.

1 I woke up one morning and discovered that I was doing feminist cultural studies of science and technology

My title refers to the Joni Mitchell song *Chelsea Morning*. It is also an ironic take on the discovery narratives of 'great men' in the history of science. More autobiographically it refers to my own relation to the coming together of a set of critical analytical activities under the label 'feminist cultural studies of science and technology' in recent years. There is a sense in which I do feel magically relabelled and relocated into a burgeoning new field. When I first started to think about cultural studies of science and technology (in the early 1980s), there were few who would have claimed that this was what they were doing. Now, it seems as if, as the cliché goes, 'everyone is doing it!' Certainly I do still retain a sense of wonder about being part of an interesting movement which has, in many respects, transformed understandings of science and technology and provided new ways of analysing their development and significance. My hope is that this book will convey some of that wonder – the excitement – around this new field, while proffering some concrete specification about what cultural studies of science and technology has to offer.

On the other hand, my suspicions about discovery narratives and stories about the spontaneous emergence of academic/intellectual fields are similar. They both effect historical condensations, making it difficult to see the complex processes involved in making both science and the academic fields which study it. Indeed, I was amused on one occasion in the late 1990s to find myself introduced as a speaker who had studied 'cultural studies of science' in Cambridge in the 1970s. Amusing though this was, it is important to remember that the self-conscious designation of the field of cultural studies of science and technology is recent and cannot be taken for granted, nor imposed anachronistically. Generally, although social studies of science and technology studies have been concerned with reflexivity and symmetry, there has been relatively little attention given to the history of the discipline (or interdiscipline) itself.[1] This book derives from my interest as a participant in that history. It reflects my curiosity about the history of the practices and perspectives which have constituted cultural studies of science over the past two decades. Thus, I try to identify and elucidate some of the practices and insights which can be clustered under the label 'cultural studies of science and technology'.

Discovery narratives, in their preoccupation with heroes, genius and magical moments, are rather monolithic origin stories which deny the work and obscure the contributions of a wide range of actors. In contrast, the impulse for this volume is much more genealogical and wide ranging. I have set out to identify diverse contributors and to trace a variety of developments that have been part of the making of cultural studies of science and technology. While I have been concerned to be reflective and analytical about distinctive features of this work, my intention is not to be prescriptive. In this sense, this is a book concerned with how work has been done but it is not a methodology text. Likewise, this is not a book about theory, although I develop theoretical scaffolding for my research projects and engage with a broad repertoire of theories in undertaking these. Rather, this book attempts to investigate some of the working toward and working through of cultural studies of science and technology.

Different versions of cultural studies of science and technology

As I have indicated, until fairly recently cultural studies and science studies were not commonly linked. The proceedings of a conference held at the University of Illinois in Champagne, Illinois, resulted in a key text – *Cultural Studies* – edited by Larry Grossberg, Cary Nelson and Paula Treichler, with a cluster of articles on 'Science, culture and the ecosystem' (Nelson, Treichler and Grossberg 1992: 21). A 1994 conference at the Center for Cultural Studies at the City University of New York Graduate Center was much more specifically focused on science and technology. Indeed, the volume resulting from this conference, *Technoscience and Cyberculture*, opens with nothing less than 'A manifesto on "The cultural study of science and technology"' (Menser and Aronowitz 1996). A few courses and academic programmes have been labelled 'cultural studies of science and technology'.[2] These are some markers of the field which do not, in themselves, constitute founding moments.

Meanwhile, definitions of cultural studies of science have proliferated and range from vague ones which could embrace virtually any form of science studies to sweeping epistemological claims. Joseph Rouse chose to define 'the term broadly' 'to include various investigations of the practices through which scientific knowledge is articulated and maintained in specific cultural contexts, and translated and extended into new contexts' (Rouse 1992: 2). The distinctive element in this definition is Rouse's highlighting of the work – 'the practices' – required in the articulation of scientific knowledge, emphasizing its context dependency. Nevertheless, his definition is, as he notes, broad.[3]

In contrast, the introductory article ('manifesto') in *Technoscience and Cyberculture* by Michael Menser and Stanley Aronowitz is quite dramatic in its declaration that cultural studies is 'the name we give to the transformation of social and cultural knowledge in the wake of an epochal shift in the character of life and thought whose origins and contours we only dimly perceive' (Menser and Aronowitz 1996: 16). In an earlier book, Aronowitz was even more sweeping in his claims for cultural studies, heralding it as the new paradigm for knowledge, which emphasizes (in contrast to the natural sciences) the contextuality of all knowledge claims (Aronowitz 1993: Ch. 7). In this same volume, an established (and now deceased) science studies researcher, Dorothy Nelkin, observed that recently some humanities researchers and social scientists have taken to 'defining their work as cultural studies of science and bringing to bear their skills in interpreting narratives and discourses' (Nelkin 1996: 34).

Hence, even in this one collection there are very different versions of *doing* cultural studies of science: from a rather modest 'add literary techniques to social studies of science and stir' approach (Nelkin 1996) to a manifesto vision of an epistemological revolution (Menser and Aronowitz 1996). The three recent definitions of cultural studies of science considered here are indicative of the instability of and diversity within this field. Nevertheless, I have chosen them because they illustrate three key dimensions of cultural studies of science and technology: epistemology, methodology and disciplinarity/trans-disciplinarity. These are some, but by no means the only, issues about this field that will feature in the chapters that follow. The preceding definitional forays came from academics working within different academic locations and traditions. As this indicates, cultural studies of science and technology has not neatly sauntered forth from cultural studies departments. Indeed, as I shall suggest in Chapter 2, science and technology studies did not feature prominently in British cultural studies of the 1970s and 1980s (see also McNeil and Franklin 1991; Reinel 1999).

My own autobiographical trajectory took me from the study of history of science as a student in Canada and Britain in the 1970s, into feminist science studies from the late 1970s, and the melding of my work, from the 1980s to the present, into what I would now designate as feminist cultural studies of technoscience. Thus, much of the research undertaken for this book reflects both my own trajectory and the adaptation and transformation of various cultural studies traditions in the study of science and technology. However, this book also registers the hybridity of cultural studies of science and technology as a field. For this field has been shaped by researchers and teachers from many disciplinary and interdisciplinary locations and traditions. Cultural studies is itself a fragmented interdisciplinary and, some would say, post-disciplinary, field. Indeed, Aronowitz (1993) advocates that cultural studies should be 'anti-disciplinary'. Nevertheless, cultural studies of science and technology has also been fabricated by the contributions of those doing cultural studies within more traditional disciplinary locations: including, for example, English, various other languages and literatures, American and other regional studies. In addition, cultural anthropology and media and communication studies have been significant locations for the development of new modes of research in the social studies of science and technology. Hence, cultural studies of science and technology is a rather ambiguous label in terms of disciplinary locations, affiliations and identities. Moreover, academic disciplines, with their particular traditions and practices, have by no means been the only influence in the forging of cultural studies of science and technology.[4]

It is not my intention to adjudicate definitions here or elsewhere in this book. I use them to register both the uncertainty about the *doing* of cultural studies of science and technology and the profound questioning associated with it. Instead, this monograph investigates, reviews and critically evaluates the distinctive features of this comparatively new field of scholarship, whilst *doing* cultural studies of science and technology. I offer my own set of explorations and enactments of cultural studies of science and technology. These take the form of a genealogical tracing of some key moments in the emergence of the field (Chapter 2) and a set of six research projects conceived as contributions to this new field (Chapters 3–8).

Feminist cultural studies of science and technology

My work within the field of cultural studies of science and technology has been further specified as *feminist* cultural studies of technoscience. This designation registers the key influence on my work within this field and the orientation of this volume. One way of describing this field would be to characterize it as marking the coming together of three relatively new, interdisciplinary fields: feminist studies, science and technology studies, and cultural studies. Evelyn Fox Keller ([1987] 1999) has described the emergence of science studies and feminist studies as two 'new' and 'parallel' fields of study in the late twentieth century. Writing in the late 1980s, she explained that

> until quite recently there has been virtually no intersection between the two disciplines, just as there has been virtually no interaction between attempts to reconceptualize gender and science – as if the two categories were independent, each having nothing to do with the other. It is only with the emergence of a modern feminist critique of science that the categories of gender and science have come to be seen as intertwined, and accordingly, that the two subjects (feminist studies and science studies) have begun to converge.
>
> (Keller 1999: 235)

Subsequently, Nina Lykke (2002, forthcoming) undertook a cartography of the fields of cultural studies, feminist studies and science studies which resonates with aspects of my project and, in fact, a version of one chapter of this book (Chapter 2) was written in explicit dialogue with Lykke's (forthcoming) mapping.

Lykke's and Keller's influential articles have been important resources for my own research projects and I discuss both in more detail in later chapters (Chapters 2 and 4). More generally, they provide me and other readers with perceptive, challenging views of the interactions between cultural studies, feminist studies and science studies. However, this book takes a rather different approach to these interdisciplinary encounters. While both Keller and Lykke consider some of the distinctive features of these fields, their cartographic approach tends to underscore their equivalence. Although it is not possible to engage in an elaborate comparison here, it is also important to register some significant differences amongst these fields. So I shall simply highlight ways in which these new fields are *not* equivalent, since awareness of these differences has shaped my approach to these interdisciplines. Science studies has a somewhat longer history within the academy, making its presence felt through history, philosophy and some sociology of science teaching and research from the mid-1960s in the UK and North America (see Edge 1995).[5] In these terms, both cultural studies and feminist studies are newer academic 'kids on the block'. However, much more important is the fact that both cultural studies (at least in the predominant British version) and feminist studies emerged from key social and political movements of the late twentieth century. In this sense, they are formations that have developed in dialogue with, or at least with reference to, a specific political constituency outside the academy. Related to this is the explicitness of the political identities of these disciplines as they emerged in the late twentieth century.[6] Science studies as a discipline is not identified with any political movement,[7] cultural studies as a discipline has a complex and uneven history of diverse political allegiances (see Chapter 2), and feminist studies is quite clearly identified with late twentieth-century and early twenty-first century feminism. Even if these identifications are ambiguous and unpredictable in their impact, they merit attention.

While Keller (1999) addresses the intersections between feminist studies and science studies and Lykke (2002, 2007) extends this to include cultural studies, there have also been some notable examinations of the encounter between feminism and cultural studies that have influenced my own book project. The first collection marking this coming-together was *Women Take Issue* (1978), produced by the Women's Studies Group at the University of Birmingham Centre for Contemporary Cultural Studies. The follow-up collection, *Off-Centre: feminism and cultural studies* (Franklin *et al.* 1991b) includes an introductory editorial overview of 'Feminism and cultural studies: pasts, presents, futures' in which Sarah Franklin, Celia Lury and Jackie Stacey set out 'to highlight' what they consider to be 'some of the key issues' in bringing together these two worlds, indicating both 'overlaps' and 'lack of overlaps' (Franklin *et al.* 1991a). Sue Thornham's *Feminist Theory and Cultural Studies* (2000) offers another interpretation of the meeting of these traditions. These studies of the productive hybridity and tensions between these two political and intellectual traditions that flourished in the last decades of the twentieth century have influenced my efforts in this volume to track and evaluate other patterns of intellectual development. Indeed, as a teacher and researcher at the Centre for Contemporary Cultural Studies (1980–96) and as a contributor to *Off-Centre: feminism and cultural studies* I was part of this interaction. Of course, these overviews are primarily concerned with the encounter between feminist studies and cultural studies. Nevertheless, both *Off-Centre* and *Feminist Theory and Cultural Studies* also bring together feminist cultural studies and science and technology studies, which I shall consider in more detail in Chapter 2.[8]

The framing of this volume as *feminist* cultural studies of science is designed to foreground the absent presence that has haunted both modern science and, even regrettably, much of the field Keller (1999) designates as 'contemporary science studies'. In the wake of second-wave feminism, a strong body of research and scholarship emerged which documented the exclusion of women from and their marginalization within the history of modern Western science. Despite the myriad interventions and extensive reforms realized since these 'revelations', the world of science has not totally thrown off its identity as 'a world without women': in that it remains a world in which men dominate and which is often coded as 'masculine'. Given the global significance of this basic, lingering gender pattern, cultural studies of science can scarcely avoid this issue. So this is the most fundamental reason for explicitly approaching cultural studies through the prism of feminism.

In undertaking my exploration of *feminist* cultural studies of science I was also responding to Donna Haraway's persistent whispers (and sometimes shouts) about absent presences in a rather different sense. I was mindful of some pointed gibes and trenchant criticisms which have peppered Haraway's writings over the past quarter of a century. In challenging but collegial comments, some of which are hidden away in her footnotes, she chides science studies researchers for their failure to engage with feminist scholarship (Haraway 1997: 26–8). She sometimes reprimands these researchers for ignoring, dismissing or underestimating particular feminist contributions and insights. Beyond these specific commentaries, writing in the early 1990s she angrily observed 'the abject failure of the social studies of science as an organized discourse to take account of the last twenty years of feminist enquiry' (Haraway 1992: 332). Mindful of this 'abject failure', in my own science studies research I was eager both to trace, in some detail, some of the moments in the coming together of feminism and science studies and to explore the rich insights feminism has provided for the analysis of science.

But foregrounding feminism in this volume was not just a matter of political principle or intellectual strategy. Framing feminist theory as a 'hermeneutical tradition', Sarah Franklin, Celia Lury and Jackie Stacey assert that there is 'no topics or phenomena to which a feminist analysis is not relevant' (Franklin *et al.* 2000a: 6). While I would endorse their assessment, there was a more immediate empirical pull towards feminism within this book project. There was a strange way in which feminism and feminists, and, in some cases, women's bodies, emerged empirically and somewhat uncannily as absent presences in the various investigations undertaken in this volume. Thus, I found it impossible to analyse the making of scientific heroes, the narratives of modern technoscientific reproduction, the technological spectacles of war and tourism without investigating this haunting.

The scope and structure of this book

Although each of these chapters could stand alone, as they emerge from distinct research projects, they contribute to my broader remit to explore the contours, orientations and significance of the recently constituted field (or sub-field) of cultural studies of science and technology. Thus, through the genealogical and six other specific studies presented in this volume I seek to highlight and to analyse distinctive features of the 'cultural turn' in science studies which has occurred during the last two decades. Across these seven chapters I interrogate the theoretical and methodological implications of this turn in science studies by considering the work of other scholars and through developing my own original case studies. Probing the field entails investigating in what sense and in what ways cultural studies of science has constituted a distinctive mode of science studies. At some points this brings me to considerations of how it might differ from more established forms of science studies, including some versions of the history of science.

Inevitably, in raising questions about how science and technology studies might be done I am also raising questions about science itself. My proposals and those that I analyse offered by other cultural studies scholars who advocate reorientations of the field are not so much academic twiggings; rather, they are designed to equip researchers for a fuller and more complex appreciation of science in contemporary Western culture. So, for example, like Roger Cooter and Steven Pumphrey (1994), I suspect that understandings of science would be transformed if the field really was fully open to the investigation of science and technology in popular culture. In addition, my explorations of and within cultural studies of science and technology begin to move beyond the investigation of cognitive relationships to include attention to pleasure, dreams, desires and other aspects of how we live science and technology which have been relatively neglected in other versions of science studies.

A further objective is to situate this eruption within science studies on a wide canvas. Hence, I consider the intellectual and political significance of this developing body of work with reference to some contentious debates about science and technology in contemporary Western societies, particularly in the United Kingdom, the United States and Canada. Notions of 'the risk society' (Beck 1992), 'public understanding of science' and 'public engagement with science' have figured prominently in such debates and they have had considerable resonance inside and outside the academy at the end of the twentieth and beginning of the twenty-first centuries. They hint at the quotidian and pervasive implication of science and technology into all aspects of contemporary lives in Western societies. They indicate that grappling with science and technology has become an issue of prime importance in these settings and they register that this entails many political challenges.

The next chapter of this book (Chapter 2) offers a genealogy of feminist cultural studies of science and technology as it emerged in Britain and North America during the last decades of the twentieth century. The investigation which informed this chapter was inspired by the burgeoning of the new field of cultural studies of science and technology and my interest in tracing routes into this work (McNeil forthcoming). More particularly, I was motivated by my firm sense that feminists had been important movers and shakers in this development. As indicated previously, I was stirred by Donna Haraway's reprimands about science studies' neglect of the challenges and promise of feminism. Hence, I set out to trace the dispersed but powerful encounters between feminism and science studies in a diverse set of projects that constituted the early stirrings of cultural studies of science and technology in the last two decades.

Chapters 3–8 present a set of self-contained explorations of science and technology in late twentieth- and early twenty-first-century Western popular culture, based on autonomous, but interrelated, research projects which I have undertaken during the past twenty years. These are organized into three sections which revolve around key tropes in the casting of modern science and technology: making heroes, telling stories and witnessing spectacles. I highlight and work through these frames because they are important strands in the making of modern science and technology. Taking these tropes seriously undermines the conventions of one of the most powerful origin stories in the Western world: that science is made by individual scientists and groups of scientists and that it radiates out from them to the rest of the world. Mark Erickson sees the related predilection to 'prioritize and fetishize science in certain locations' in the assumption that 'we "know" where real science is – in the laboratory, the textbook, the documentary' (Erickson 2005: 24) as a manifestation of scientism. Likewise, Bruno Latour insists that the 'few people officially called "academic scientists" were only a tiny group among the armies of people who do science' (Latour 1987: 173). But, as I shall argue in this book, this fixation with particular agents and locations has also cast its shadow over much of science studies itself. For this reason Donna Haraway cautions that technoscience 'should not be narrated or engaged only from the point of view of those called scientists and engineers' (Haraway 1997: 50). Focusing on the explictly public processes of making heroes, telling tales of technoscience and witnessing technoscientific spectacle is thus a deliberate strategy to shift attention to a wider range of agents and locations implicated in the making of science and technology. Pursuing these tropes registers the multiplicity of cultural processes involved in the making of modern science. I have used these frames to structure this book because they designate quotidian processes through which members of the public in the Western world live with and make science and technology in their daily lives. Analysing their operations highlights some of the ways science and technology are made in everyday life and in popular culture. Hence, this tropic analytical structure underscores my contention that science is not simply what scientists do, say or write.

There are two chapters within each of the three tropic sections: making heroes, telling stories, witnessing spectacles. Each of the chapters constitutes a discrete study of science and technology in recent Western industrial culture. Chapter 3 presents and examines a set of case studies about Newton as a celebrated hero in British culture from the end of the eighteenth century through to the contemporary period. Chapter 4 explores the writing of scientific heroism beginning with James Watson's controversial memoir *The Double Helix* ([1968] 1969) and moves on to examine three of the most popular biographies of women scientists to emerge in the 1970s and 1980s: Anne Sayre (1975) *Rosalind*

Franklin &DNA, June Goodfield ([1981] 1982) *An Imagined World: a story of scientific discovery* and Evelyn Fox Keller (1985) *A Feeling for the Organism: the life and work of Barbara McClintock*. Chapter 5 uses autobiographical and dream narratives to investigate the patterns of storytelling around and the political significance of the emergence of in vitro fertilization (IVF) in the 1980s and early 1990s. Chapter 6 follows on from the preceding chapter, tracing the development and changes in popular narratives around reproduction and their increasing significance in the emergence of 'intimate citizenship' (Plummer 1995) in Britain and other Western countries at the turn of the twenty-first century. Chapter 7 analyses the Gulf War of 1991 as a crucial episode in the genealogy of military technoscience and technoscientific spectacle. Focusing on television coverage, popular Hollywood film and video (and computer) games, it considers how technology fuelled 'banal nationalism' (Billig 1995), the instantiation of a 'new world order' and corporeal disappearance. Chapter 8 examines technological spectacle and feminist viewing and transformative views, focusing on two main US tourist sites – Disney's EPCOT Center (Orlando, Florida) and the NASA Spaceport at Cape Canaveral, Florida.

Technoscience studies

The term technoscience is employed throughout this volume. As Donna Haraway notes, this term 'communicates the promiscuously fused and transgenic quality of its domains' (Haraway 1997: 4). She contends that '[t]his condensed signifier mimes the explosion of science and technology into each other in the past two hundred years around the world' (Haraway 1997: 50). Haraway attributes the 'common adoption' of the term within science studies to Bruno Latour (Latour 1987; Haraway 1997: 279, n. 1).[9] The term technoscience is thus a critical concept which highlights this fusing, but as Haraway (1997: 51), Sismondi (2004) and Erickson (2005: 10) observe, it has also become a shorthand way of acknowledging how science has come to pervade all aspects of contemporary life. In this sense, it evokes the technologized state of late twentieth- and early twenty-first-century everyday life in Western societies.

I support the adoption of the term technoscience and subscribe to the critical perceptions which inform its use. However, I am mindful that the processes whereby science and technology became fused have been realized gradually and unevenly over the 200-year period Haraway references and that these themselves merit attention. Thus, while it is virtually impossible to draw a line between 'pure' science and its application in technology in the Western world in the early twenty-first century, we must be careful not to project this fusion as a seamless, monolithic picture of modern science from the seventeenth century to the present. Accordingly, while undertaking a study of the patterns of the modulations of science and technology is well beyond the remit of this book, I have, at some points, distinguished science from technology where this seems appropriate. I do this, for example, in my discussion of Newton's significance in the eighteenth and early nineteenth centuries (Chapter 3).

Moreover, although some have chosen to designate their research specifically as technoscience studies, the broad disciplinary labels of science studies, social studies of science and science and technology studies are widely circulated (and used interchangeably), with the latter terms encompassing research covering the entire historical period from the seventeenth century (and, indeed, sometimes earlier) to the present. My own use of these terms registers and enacts this pattern of usage. The term technoscience will be used with specific reference to science and technology in the recent and contemporary Western world – in the second half of the twentieth and early twenty-first centuries in particular, but I deliberately do not use this term with reference to much earlier periods. In referencing academic disciplinary fields I adopt the terms science studies, science and technology studies and technoscience studies according to these conventions.

This monograph investigates, reviews and critically evaluates the distinctive features of this comparatively new field of scholarship, whilst *doing* cultural studies of science and technology (investigating scientific heroes, reproductive technologies, military technology, technological tourism). The book is structured to highlight the multiple and dispersed makings of the meanings of science and technology in everyday life and popular culture. It challenges the assumption that science is simply what scientists do, say or write, thereby raising broader questions about science and technology in Western societies. The third strand of the book is its reflective component. This means that methodological and theoretical questions have been foregrounded as I consider what is involved in *doing* cultural studies of science and technology.

2 Feminist cultural studies of science and technology
Roots and routes

As I noted in Chapter 1, cultural studies of science did not magically emerge in the last decades of the twentieth century, although the recent rapid proliferation of the term might create that impression. This chapter addresses the very basic question: 'How did cultural studies of science and technology come about?' However, it tackles this question not by telling an origin story, but rather by doing some tracing: uncovering and highlighting some of the multiple makings of cultural studies of science and technology. The question about how cultural studies of science and technology became a practice must be answered with reference to feminism. As I shall show, feminist researchers have been key agents in the making and doing of cultural studies of science and technology. For some time now, feminist scholars have decried the resistance to gender analysis and feminist insights within social studies of science (see, for example, Harding 1986; Delamont 1987; H. Rose 1994; Haraway 1997: esp. 26–8; Wajcman 2000, 2004). Inspired by these protests, I set out to document in this chapter how the relatively new kind of science studies – cultural studies of science and technology – was forged and feminist researchers figure prominently in my account.

My mapping of the dispersed and diversified development of cultural studies of science and technology also entails some detailed consideration of what was involved in crafting new modes of analysis for science studies. Investigating these modes of research establishes that, while there have been marvellous moments of innovation and ingenuity in the new field of feminist cultural studies of science, these are the products of considerable work. Hence, this chapter is a charting of transpositions, translations, adaptations, transformations, as well as innovations. My intention is to celebrate the achievements and promise of the field by detailing the features of specific modes of *doing* feminist cultural studies of science and technology. What follows is not a cartography of a unified or uniform terrain. Instead, I highlight tensions and discordances, as well as common traits, as part of my reflections on the contours of the emerging field of feminist cultural studies.

The exploration of cultural studies of technoscience which is presented in this chapter is, to some extent, in dialogue with the related project undertaken by Nina Lykke (2002, forthcoming). Lykke maps feminist cultural studies of technoscience as an implosion of the overlapping fields of science studies, cultural studies and feminist studies. In contrast, this chapter is a *genealogy* – tracing patterns of emergence in some analytical practices *over time*. I set out to investigate specific developments involving feminist influence on and shaping of the emerging field of cultural studies of technoscience. I am offering here only my own 'partial perspective' (Haraway 1991b) in a genealogical account which focuses mainly on Anglo-American research. My hope is that my exposition of some versions of feminist technoscience in Anglo-American settings will encourage reflections about other paths taken into this field.

Different roots and different routes

As the preceding brief review indicates, cultural studies of technoscience means different things and may involve different activities. Nevertheless, it is possible to say something about some routes into this trans-disciplinary domain. In this chapter I consider approaches coming from five different Anglo-American disciplinary/interdisciplinary roots/routes:

1 cultural anthropology;
2 literary studies of science;
3 studies of visual culture (through art history and film studies);
4 British cultural studies;
5 feminist science fiction studies.

The use of disciplinary and interdisciplinary labels as tracking devices seems appropriate and convenient, since it helps to denaturalize the mythical assumption that all research in feminist studies, cultural studies and technoscience studies is inherently inter- or trans-disciplinary. I want to tease out trajectories towards interdisciplinarity and highlight methodological borrowings and adaptations that are sometimes obscured by this kind of naturalization. Moreover, using this taxonomy is also a reminder about institutional locations that can both support and confine researchers. Although I cannot pursue this here, it is my way of hinting at institutional struggles, contestations and transformations that have been part of these trajectories.

Like all divisions, the demarcation of these routes is limited in many ways. It overestimates the significance of academic locations rather than other factors which I shall highlight and discuss in tracing intellectual and political engagements. Such charting does not register overlaps.[1] Nevertheless, there are important distinctive features to each of these strands and they merit attention in this chapter, which also registers fruitful exchanges and cross-fertilizations between these.[2]

Route 1: cultural anthropology

Cultural anthropology has its own roots in Western imperial explorations during the nineteenth and early twentieth centuries. From the encounters between Western explorers and colonialists of this period emerged practices of systemic observation, documentation and characterization of 'other' cultures. These methodological traditions were channelled into this discipline focused on ethnography which involves multi-method immersion in cultures for a significant period and fieldwork in a 'foreign' culture. The object of such fieldwork was the production of insightful analysis of 'other' (non-Western) societies and thus 'culture' emerged as the central focus and key term of this discipline.

Cultural anthropology has been closely associated with colonialism and imperialism and their legacies. From the 1970s through the 1990s, however, an explicit, critical confrontation with these origins fostered a radical turn within the discipline which opened its resources for science studies. With an awakened awareness of the discipline's colonial links, some cultural anthropologists turned their analytical gaze in the direction of Western culture itself. Controversial though this was, the result was a fresh body of investigation and scholarship that attempted to distil and document features of Western societies and cultures.

Once Western societies and cultures had come under anthropological scrutiny and become the subject for fieldwork investigation, it was almost inevitable that the discipline would begin to examine science as a crucial element of Western culture. Nevertheless, it is important not to underestimate either the controversy surrounding this new direction of the discipline or its significance. David Schneider, Marshall Shalins and a few other established figures led the discipline in this direction. However, feminist scholars were also in the vanguard of this new branch of anthropological study. Among these scholars, some, including Margaret Lock, Marilyn Strathern and Helen Verran, began their careers doing fieldwork in the traditional anthropological manner – observing non-Western cultures. Their earlier fieldwork experiences informed their subsequent investigations of Western science and medicine. Other feminist anthropologists, mainly from the 'next generations' in the discipline, did their fieldwork in the West exclusively.

Feminist scholars have adapted two key anthropological methodological and theoretical traditions in anthropology – ethnographic research and kinship studies (sometimes combining the two) – in order to investigate science and technology in Western culture. The rich tradition of ethnographic study has been brought back 'home' by anthropologists and others who have undertaken recent studies of the daily cultures of laboratories, clinics and other scientific settings. This strand of feminist cultural studies undertook the investigation of rituals and practices instantiated by and integral to technoscience.

Nevertheless, it is feminist work on reproduction and kinship that has been most prominent in this branch of cultural studies of technoscience. Engaged with reproductive politics through feminism, and faced with the development of new reproductive technologies in the late twentieth and early twenty-first centuries, some feminist anthropologists turned kinship studies in a new, critical direction. The research of Jeannette Edwards, Sarah Franklin, Faye Ginsburg, Emily Martin, Rayna Rapp and others was exemplary in realizing this reorientation. Davis-Floyd and Dumit's, *Cyborg Babies: from techno-sex to techno-tots* (1998) is a collection which demonstrates the rich vein of feminist studies of reproductive science and technology along this anthropological route.

At the heart of much of the feminist work in this stream of anthropologically oriented science and technology studies has been an ambivalence about the 'othering' of cultures that is characteristic of anthropological study. As previously indicated, the acknowledgement of anthropology's entanglement with colonialism and imperialism brought a suspicion about its role in 'othering' non-Western cultures. On the one hand, the turn to Western science and technology constituted a disavowal of this distancing, bringing anthropology home, as it were. On the other hand, one of the strengths of anthropology was that it tended to render strange – 'other' – precisely those features of Western culture (including science) which are often naturalized. Moreover, critical perspectives, including those associated with feminism, generated a critical distancing from Western science within studies of science and technology. Interestingly, such 'othering' of science and medicine has been problematized in some recent critical technoscience studies research (Mol 2002).

Feminist anthropological science studies emerged as part of a reflexive turn in the discipline itself. This turn was informed and shaped by encounters with postmodernism and poststructuralism. One manifestation of this turn was greater exploration of writing as a key part of anthropological practice (Clifford and Marcus 1986). This was also a feature of the intensified analytical and ethical concern about the role of the investigator. Feminist scholars have been at the forefront of this development, but they have also been involved in gingering up critical perspectives on ethnography as a rich, but problematic, methodological tool chest that continues to be unpacked in trying to understand and transform the power relations of the discipline.

Route 2: literary studies of science

As already indicated, feminist scholars and others turned cultural anthropology on its head in undertaking their cultural studies of science and technology. In ethnography and the tradition of kinship studies they found complex and rewarding sets of research methods and frameworks. Feminist literary scholars brought to science studies a rather different, but equally rich, repertoire of practices culled from literary and linguistic disciplines. This strand of feminist cultural studies of science was first forged through the scrutiny of scientific texts as cultural artefacts that could be subjected to literary and linguistic analysis. Although initially such approaches were cautiously framed, this research firmly identified science as cultural.

The appearance of Gillian Beer's *Darwin's Plots: evolutionary narrative in Darwin, George Eliot, and nineteenth-century fiction* ([1983] 2000) was a key moment in the emergence of this strand of cultural studies of science. Three years later came a related anthology, edited by Ludmilla Jordanova, *Languages of Nature: critical essays on science and literature* (Jordanova 1986). Taken together, these two volumes assert the textuality of science, locate scientific writing in a broad panorama of cultural forms, and subject scientific texts to diverse forms of literary and linguistic analysis.

Beer (2000) unequivocally brings literary studies into the domain of science studies in a book that has been described as 'deliberately "literary" in its approach' and in which she 'sets out to read Darwin as a writer who also happened to be a scientist' (Levine 2000: xi). By locating *On the Origin of Species* (1859) on a complex cultural tapestry, Beer establishes that Darwin's evolutionary theory can be fully deciphered and appreciated only through reference to poetry and fiction. She argues that 'how Darwin said things was a crucial part of his struggle to think things, not a layer that can be skimmed off without loss' (Beer 2000: xxv). Tracing Darwin's 'metaphors, myths, and narrative patterns' (Beer 2000: 5), Beer shows the cultural resources, particularly those deriving from an English literary heritage, from which Darwin drew in building his arguments.

Beer's work was a key turning point in the emergence of cultural studies of science by way of literary studies. For some, this was a 'bridge too far' between the 'two cultures' (arts and sciences) which C. P. Snow had identified in the 1950s (Snow [1959] 1993). This is obvious from George Levine's foreword to the second edition of the book, in which he praises Beer, while warding off the threat of Darwin's theories being seen as merely cultural. He insists that Beer has refused 'historical or social reductionism' and maintained Darwin's 'special genius' and that her study of the language of Darwin's texts 'undercuts the implication that Darwin was absolutely a man of his time, explicable in terms of the conventions of the middle-class society to which he so nervously and doggedly adhered' (Levine 2000: xi, xi, x).

Despite the many achievements of Beer's book, there are also strictures on her version of cultural studies of science. Her adoption of a rather Kuhnian (Kuhn [1962] 1996) framework in conceptualizing scientific theories imposes some temporal contingency on the contention that science is fully cultural. The comment in the opening paragraph of *Darwin's Plots* is indicative in this respect: 'When it is first advanced, [scientific] theory is at its most fictive' (Beer 2000: 1; my interjection). Here, Beer seems to suggest that scientific theory gradually loses its cultural moorings or associations.

Moreover, the book vividly conveys Beer's regard for Darwin as a creative writer and her sense of his genius – his singularity. The emphasis on genius, singularity and high culture countervails and contains the promise of providing a full cultural study of Darwin's theory. Hence it is interesting to find Beer explicitly distancing her project from that of Charles Darwin's biographers, Adrian Desmond and James Moore (Desmond and Moore 1991), who present Darwin as embodying and exemplifying nineteenth-century, middle-class English masculinity. In fact, whereas Beer traces the links between *On the Origin of the Species* and the established English literary canon, Desmond (1984) investigates the connections between Darwinian scientific theory and popular culture.

While her original study (Beer [1983] 2000) involves a fairly traditional literary and high-culture version of cultural studies without reference to gender relations, sexuality or feminist perspectives on science, Beer subsequently linked her work to a new wave of scholarship in feminist science studies. In her preface to the second edition of *Darwin's Plot's* (Beer [1983] 2000) she praises feminist science studies, which she presents as akin to her own earlier project. Although the second edition of her book is not a revised edition, Beer does wistfully indicate her wish that she could have drawn on that scholarship in her own study. Between the first and second editions of Beer's landmark text, other feminists had used and extended literary frameworks in diverse directions. The volume edited by Mary Jacobus, Evelyn Fox Keller and Sally Shuttleworth, *Body/Politics* (Jacobus *et al.* 1990) is perhaps the best-known publication of this sort, although it is much more interdisciplinary than Beer's study. Discourse is its linking concept, the female body in biomedical discourses is its predominant focus, and psychoanalysis figures in a number of the contributions to this explicitly feminist collection.

Nina Lykke (forthcoming) has registered the importance of the International Society for Literature and Science (founded in 1985) in the emergence of feminist cultural studies of science. During the 1980s and into the 1990s Beer and Jordanova were leading figures within the emerging interdisciplinary field this society represented. Debates about the relationship of the study of literature and science to traditional literary studies continued and the question of whether such research should embrace explorations of popular culture loomed symptomatically over the field. For example, Lykke notes the debates within the International Society for Literature and Science about relabelling itself and broadening its remit. She reports the Society's decision to replace the title 'literature' with 'art' (signalling its associations with elite culture) rather than 'culture' (which would have included popular culture within its remit).

Route 3: studies of visual culture (through art history and film studies)

The literary orientation of the strand of cultural studies of science forged by Beer ([1983] 2000) and Jordanova (1986) is striking. The next major project by Ludmilla Jordanova (1989) – *Sexual Visions: images of gender and medicine between the eighteenth and twentieth centuries* – sets out in a rather different direction. In that work, Jordanova embarks explicitly on gender and sexuality studies, informed by feminist scholarship, but she also unleashes cultural studies of science from its exclusively literary or linguistic moorings. In addition to exploring literary texts, she analyses paintings, sculptures and scientific diagrams to provide an interesting set of studies of gender relations in biomedical science between the eighteenth and twentieth centuries. By using the analytical repertoire of art history to bring science's visual dimensions into cultural studies of science, Jordanova complexifies understandings of the cultural dimensions of science. She categorically and methodologically extends and enriches cultural studies of science as the study of high visual culture.

Ludmilla Jordanova's explorations of the visual dimensions and resonances of science influenced a group of feminist scholars working in film studies, who shaped a distinctive trajectory in cultural studies of science. Two special issues (nos. 28 and 29) of the feminist film studies journal *camera obscura*, both titled 'Imaging technologies, inscribing science' (Treichler and Cartwright 1992a, 1992b), effectively announced this new sub-field and assembled exemplar studies. These volumes were edited by Paula A. Treichler and Lisa Cartwright but the contributors included Ann Balsamo, Giuliana Bruno, Valerie Hartouni, Constance Penley, Ella Shohat and Carol Stabile, who all became associated with feminist cultural studies of technoscience.

There were a number of distinctive features of this strand of feminist cultural studies of science. The two special issues of *camera obscura* had been inspired by political activism, and the editors regarded their research as a political intervention in its own right. The introduction linked these journal issues with 'a new wave of activism toward institutionalized science and medicine' (Treichler and Cartwright 1992a: 5) that was associated particularly with groups of AIDS activists (such as ACT UP) and the National Black Women's Health Project in the USA. They reflected that such groups' 'highly visual and theatrical activist strategies articulate and publicize important critiques of science and medicine' (Treichler and Cartwright 1992a: 5).

This explicitly political orientation informed and shaped the research presented in these two issues of *camera obscura*. The pivot of much of the research was the identification of imaging that was common to science and popular film, such as the demonstration of links between the clinical or scientific gaze and the cinematic gaze. The editors observed: 'The technological history and culture of film and television repeatedly overlap and intersect with developments in scientific imaging; science, in turn, uses popular imaging conventions even as it remains emphatically aloof' (Treichler and Cartwright 1992a: 6). Moreover, the contributors to these issues of the journal were also mindful that 'many imaging technologies are targeted and have special implications for women' (Treichler and Cartwright 1992a: 8) and that investigations of such imaging would necessarily entail gender analyses.

These special issues of *camera obscura* contained articles that became, in effect, trailers for books published subsequently by Cartwright (1995) and Bruno (1993), who have been important figures in bringing film studies into cultural studies of science. In rather different but complementary projects, Cartwright (1995) and Bruno (1993) explored the intersections between the genealogies of physiology and anatomy and the emergence of cinema. In both cases, the author identifies the female body as a key focus in the development of these different and powerful cultural technologies.

Cartwright (1995) undertakes a set of detailed case studies of the use of visioning technologies in the development of early twentieth-century physiology. Bruno (1993) investigates the emergence of popular cinema in early twentieth-century Naples, arguing that its forms of spectatorship derive directly from the practices of anatomical science. Writing with reference to *La neuropatologia*, a film made by Dr Camillo Negro and first shown in Turin in 1908, she observes:

18 *Feminist cultural studies of science*

> It is the very nature of the film apparatus, with its inscription of spectatorship, that makes possible the reenactment of the anatomy lesson. ... The space of cinema replaces the geography of gazes, the interaction of subject/object, the topography of the spatio-visual representation that belonged to the medical spectacle. The vertical anatomic table becomes a white film screen, and it is the anatomy of female hysteria that is performed and exhibited.
>
> (Bruno 1993: 74–5)

She regards psychoanalysis as a new human science that emerged from the replacement of anatomical performance and the popular focus on the figure of the hysterical female body:

> The audience who customarily attends the performance of the anatomy lesson or watches the theater of hysteria becomes institutionalized as spectatorship. ... *La neuropatologia* transforms the anatomic table into both filmic screen and psychoanalytic couch and points to the female body as the ground upon which this transformation is enacted.
>
> (Bruno 1993: 75)

Cartwright, Bruno and others, noticing the common focus on the body, particularly the female body, that bonds the biological sciences, scientific medicine and psychoanalysis to cinema, and observing the importance of the use of visual technologies in the development of physiology and anatomy, were launched on investigations into what they labelled the 'common visual language of these fields' (Treichler and Cartwright 1992a: 5). They also noted the increasing use and importance of visualizing technologies in scientific and medical practices. Hence, the centrality of media was registered and they were investigated to enrich understandings of the cultural dimensions of science. Moreover, feminist analyses of the female body as spectacle became important in this strand of cultural studies of technoscience.

Route 4: British cultural studies

Cultural studies emerged as a distinctive interdisciplinary field of research and education in Britain in the 1970s and 1980s (Johnson 1983; Turner 2003). The Centre for Contemporary Cultural Studies at the University of Birmingham was a key institutional site for this development, which drew on the disciplines of English literature, sociology and history in studies of the making of meaning in everyday life in 1970s Britain. Linked originally to the New Left politics of the 1960s and 1970s, from the 1970s to the 1990s British cultural studies was challenged and, to some extent, reconfigured in the wake of feminist and anti-racist activism and research. Moreover, it was enriched methodologically and theoretically through encounters with structuralist theory (particularly that of Saussure and Barthes) and Marxism (especially through the work of Althusser and

Gramsci), and in later incarnations it adapted or responded to poststructuralist and postmodernist theories.

The literary orientation of early work within British cultural studies established a tradition which was surprisingly respectful of the English division between the 'two cultures' (Snow [1959] 1963) of the arts/humanities and the natural sciences. Moreover, there were other factors that militated against the possibility that science and technology studies might come under the scrutiny of cultural studies researchers in Britain during those years. Marxist scientism cast its shadow strongly in this field. Louis Althusser was one of the main theoretical influences in cultural studies during the 1980s, and his contention that there had been an epistemological break between Karl Marx's pre-scientific and his scientific writing in *Capital* was indicative and influential. More generally, the preoccupation with popular and alternative culture(s) among these cultural studies researchers distracted them from the study of science, which was identified with 'high' culture. Indeed, in a recent study of research methodologies and the tradition of ethnographic research in British cultural studies, Ann Gray (2003) has shown an effective aversion to 'researching up' which also contributed to the neglect of the world of science.

The interest in developing fresh perspectives on popular culture, and on positively investigating active relationships to the mass media, also estranged many cultural studies scholars from the main critical, 'continental' theoretical tradition which did focus on science – the Frankfurt School. In fact, in many respects, British cultural studies became antithetical to the perspectives of the Frankfurt School because that group was identified with negative evaluations of popular culture, associated with the emergence of mass media. This foreclosed the possibility of engagement with its critical perspectives on the legacies of the Enlightenment, scientific rationality and technological progress.

Raymond Williams was the only 'founding father' and one of the few prominent figures in British cultural studies who investigated scientific and technological dimensions of popular culture. Williams' fascinating genealogy of the 'Ideas of nature' (R. Williams [1972] 1980) in British culture and his investigation of the shifting boundaries between nature and culture in *The Country and the City* (R. Williams 1973) were somewhat anomalous, but nevertheless influential, projects in British cultural studies. Raymond Williams (1974, 1989) also argued forcefully that researchers in media studies should register and explore the technological dimensions of the mass media.

By the late 1980s, the neglect of science and technology appeared as a glaring lacuna in British cultural studies, one which feminists and others were keen to redress (McNeil and Franklin 1991; Reinel 1999). In the wake of the feminist 'body politics' of the 1970s and 1980s (Thornham 2000: ch. 7), critical investigations of the constitution of 'the natural', and of the operation of scientific and medical discourses, appeared crucial to understandings of the everyday life of women (and men) in Western societies. The winds of theoretical change ushered in by poststructuralism and postmodernism seemed to be blowing in a similar direction, given that ideas of progress and the grand narratives of science were brought under scrutiny by these theoretical movements.[3]

As Althusser's hold waned, Foucault became a more influential figure within British cultural studies. Foucault's interests in the clinic, biopower, and in various aspects of bodily discipline and medical and scientific discourses provided the theoretical ballast for new kinds of feminist cultural studies of science and technology. Outside the academy, social and political controversy that was focused on technology, science and medicine was at the centre of new social movements, such as the anti-nuclear and ecology movements, as were campaigns around HIV/AIDS. As the field of British cultural studies was reshaped during the 1980s by feminism, black British anti-racism and sexual politics, science and medicine loomed large in the mapping of forms of sexism and racism (Science and Technology Subgroup 1991; Thornham 2000; M. Barker 1981). Moreover, researchers attuned to the changes in daily life in the Western world observed the rapid developments in communication and information technologies as well as in the so-called new reproductive technologies in the last decades of the twentieth century (see Chapters 5 and 6).

Changing political priorities, shifting theoretical orientations and technological developments drew some cultural studies scholars toward the study of science and technology in the late 1980s and 1990s. In the United Kingdom and the United States, this shift was manifested in a cluster of case studies of science and technology in popular culture which appeared in the early 1990s. In the late 1980s, one of the study groups at the Centre for Contemporary Cultural Studies mapped the methodological and theoretical terrain for such interdisciplinary work (McNeil and Franklin 1991). This group of feminist researchers also analysed a contemporary episode in abortion politics in the United Kingdom: the controversy around the Alton Bill, which had been introduced into the British parliament in the autumn of 1987 with the aim of shortening the time during which abortion could be performed legally. The group, having worked 'collectively ... [on] examining the place of science and technology within cultural studies', approached this case study with a 'combination of approaches from feminist and cultural studies' (Science and Technology Subgroup 1991: 147, 148). In addition, Constance Penley and Andrew Ross (1991b) published a collection of essays that brought together feminist studies, cultural studies and technoscience studies. In their introduction to this collection, the editors emphasize that the essays are 'almost exclusively focused on what could be called *actually existing technoculture* in Western society' (Penley and Ross 1991a: xii, their emphasis). These grounded studies were offered to counter hyperbolic claims about liberation through, or total domination by, new technology. The cultural studies imperative – to study the creation of meaning in everyday life – led many of these scholars to studies of popular and alternative cultures linked to science and technology, including campaigns around abortion rights (Science and Technology Subgroup 1991), AIDS activism (Treichler 1991) and computer hackers (Ross 1991).

The focus on 'ordinary people', social and political activists and members of particular subcultures and countercultures was continuous with established modes of doing cultural studies, but it was rather more challenging in science and technology studies. Here the locus of investigation was shifted away from scientific theories, scientists, technologists, or even scientific texts or scientific or technological artefacts. Such cultural studies of technoscience were far more interested in analysing the dispersed and diverse creations of meanings around science and technology in popular culture. The heretical implication of this new kind of technoscience studies was that scientists and technologists did not control and could not regulate the meanings of technoscience.

This strand of cultural studies highlights the notion that an understanding of technoscience necessarily involves studying the multiple, diverse and complex interactions with science and technology that mark daily life in the contemporary Western world. This part of the field is also characterized by specific methodological and theoretical adaptations of British cultural studies. One example is the employment of the concepts of subculture and counterculture. Penley and Ross (1991b), for instance, in addition to their joint work, were also undertaking individual studies of specific subcultures and countercultures. Penley (1997) analysed *Star Trek* and its followers, and Ross (1991) studied environmental groups in the contemporary United States. Research projects such as that of Penley (1997), Ross (1991) and the Science and Technology Subgroup (1991) often appeared under the rubric of 'case studies'. This may reflect some tentativeness in the wake of postmodernist and poststructuralist suspicions about grand narratives and theories. However, it also emphasized fragmentation and dispersal in patterns of meaning creation around science and technology. These methodological and theoretical features are consonant with the political orientation of this version of cultural studies of technoscience, which contends that we are all active in making meaning in contemporary science and technology.

Route 5: feminist science fiction studies

Second-wave Western feminism generated a distinctive body of imaginative writing that conjured and explored encounters with technoscience. Feminist science fiction, which appeared in the wake of the women's movement of the 1960s and 1970s and of the gender and sexual politics of the last decades of the twentieth century, bent and transformed this established genre of popular literature (Barr 1981; Rosinsky 1984; Lefanu 1988; Armitt 1990; Roberts 1993; H. Rose 1994: ch. 9; Donawerth and Kolmerten 1994; Donawerth 1997).[4] Indeed, since Mary Shelley's *Frankenstein* (1819) is taken to be one of the earliest science fiction (SF) texts, feminism's 'golden age of SF' (H. Rose 1994: 209), in the last quarter of the twentieth century, could be viewed as a feminist reappropriation of the genre.[5] As a cultural form, science fiction was a distinctive imaginative space in which writers and readers could explore aspirations and fears pertaining to the increasing technologization of Western societies.

Donna Haraway celebrated this explosion of feminist science fiction during the late twentieth century. She registered the significance of the flourishing of this cultural form, welcoming the new feminist science fiction writers of the

1970s and 1980s as 'our story-tellers exploring what it means to be embodied in high-tech worlds' (Haraway 1991a: 173). Beyond this, Haraway incorporated science fiction into her studies, inviting the field of technoscience studies to recognize that science fiction is a crucial cultural site in the creation of meanings around science and technology. Her 'Cyborg manifesto' (Haraway [1985] 1991a) illustrates the promise of feminist science fiction as a resource for critical technoscience studies research.

Haraway herself demonstrates that complex cultural analysis is required for an understanding of the making of modern science and technology. In this respect, a very indicative achievement is her having combined studies of such diverse sites as museums, advertising, popular film, scientific texts, biographies and autobiographies in the production of her account of the emergence of the twentieth-century science of primatology in *Primate Visions* (Haraway 1989). Conceptually and methodologically, however, one of her most important and relatively neglected contributions to the field of social studies of technoscience is her having brought it into dialogue with feminist science fiction.

Others have followed Haraway's lead: bringing together feminist science fiction and critical analytical studies of technoscience in exploratory ways. Mary Flanagan and Austin Booth's collection *Reload: rethinking women + cyberculture* (Flanagan and Booth 2002) is one such project. The editors explain that the volume 'brings together women's fictional representations of cyberculture with feminist theoretical and critical investigations of gender and technoculture'. In fact, each of the individual contributions is starkly labelled as 'fiction' or 'criticism' and the editors frame their project cautiously:

> Rather than collapsing fiction and criticism or sharply separating practice and theory, this collection provides a variety of viewpoints from which to consider ... the effects of profound and rapid technological change on culture, particularly in women's lives.
>
> (Booth and Flanagan 2002: 1–2)

The editors and many of the contributors to the volume have recognized, as Haraway also does, that science fiction is 'a vital source of narratives through which we understand and represent our relationships to technology' (Flanagan and Booth 2002: 2).[6] This source has proved to be a rich resource for those working at the intersection of technoscience studies, gender studies and cultural studies.

Concluding coda

In this coda, I will offer some tentative reflections on my genealogical mapping by considering commonalities and variations in the versions of cultural studies of technoscience that I have traced, and then I will return to the issue of feminist affiliations. There are a number of common elements threading through the diverse versions of doing cultural studies of science that I have sketched in the foregoing account. Most obviously, all of the approaches I have highlighted share perceptions of science as culturally produced. As I have indicated, linking science with other cultural forms (such as popular film) or identifying and analysing specific cultural forms within science itself (such as the scientific text) has facilitated this recognition. The insistence on the cultural specificity of science has directly challenged assumptions about science as universal and transcendent. Moreover, this insistence has generally been accompanied by an increased awareness of science as the product of Western and modern or postmodern culture, albeit a product that has been transported and translated into other global locations and thereby transformed.

This attention to specificity has been accompanied by a greater sense of the complexity of the making of science among cultural studies researchers. A glance across the spectrum of research and investigation covered in this chapter makes it apparent that doing cultural studies of science can encompass the study of a wide range of actors and sites. For many of the researchers mentioned in this chapter, scientists are by no means the only agents in the making of science.

In many respects, the emergence of cultural studies of science has been a methodological explosion within science studies. As I have indicated, researchers have borrowed extensively from the methodological and theoretical tools of anthropology, literary studies, art history, film studies, British cultural studies and science fiction (and this list is by no means exhaustive). Inevitably, these borrowings have been selective. The melding of kinship studies in the investigation of new reproductive technologies is perhaps the most obvious example of this. Moreover, such studies have inevitably adapted and transformed methods and theories, including those of anthropological kinship studies.

In virtually all of the work considered in this chapter, researchers have drawn attention to the *mediated* nature of scientific knowledge and reminded us that no cultural form – including those forms in science – is transparent. As a result, the field of science studies in general has become much more inclined recently to pay attention to writing, images and other visual renderings. Unpacking the detailed forms of mediation has been a key methodological thread in the cultural turn of science studies.

While the features just cited are more or less shared elements of the various feminist cultural studies routes I have traced here, these trajectories also diverge in notable ways. As I have delineated, some of these researchers retain the traditional focus on scientists or scientific communities, taking their texts and/or artefacts as the starting point for science studies. Others are more interested in the diffused and dispersed generation of meanings around science and technology, investigating everyday, 'lay' encounters with science and technology, whether through medical treatment or ecopolitics or *Star Trek* fan groups.

In general, cultural studies has developed an awareness of the historical significance of the high-culture/low-culture divide and has sought to undermine that distinction. This preoccupation poses interesting dilemmas for those exploring the world of science, with its established high-culture affiliations. Some researchers have respected and indeed reinforced that characterization of science. As noted, Gillian Beer ([1983] 2000) exemplifies this orientation. In contrast, much in the field of feminist cultural studies of science has challenged this distinction, highlighting the multiple agents and agencies involved in the making of science in everyday life.

My broad-brush review of ways into cultural studies of technoscience has also taken account of the field's methodological and theoretical diversity. Textual analysis, visual analysis and ethnographic research are amongst the approaches that have been employed. The texts that have emerged from cultural studies of technoscience have themselves taken multiple forms, including thick description, contextual accounts, ethnographic studies and theoretical studies. Indeed, the theoretical borrowings undertaken by those working in this field have been broad and eclectic and by no means restricted to feminist frameworks.

In addition, these traditions may be distinguished in terms of whether or not they are concerned with investigating the power dimensions of technoscience. Those who have incorporated power as a dimension of their investigations have used different frameworks – Marxist, feminist, Foucauldian, poststructuralist, postcolonial or various combinations of these. Haraway's work exemplifies this, since it is critically oriented toward the power relations of technoscience, while her studies employ a range of theoretical resources drawing on all of these frameworks. Some of the work cited here is reflective, methodologically and analytically, about the position of the technsocience studies researcher. In other cases, researchers have adopted, rather than repudiated, the scientific norm of ostensible value neutrality.

Science traditionally has been seen as a framework for knowledge production, and this feature predominates in the history of social studies of science. In many respects, cultural studies of science have pushed beyond the cognitive dimensions of science. The interest in science fiction and the various detailed research projects about science and technology in everyday life tracked above (on pp. 20–2) are indicative of this push. Science fiction explores the dreams and nightmares around science and technology in the modern world. For many researchers, doing cultural studies of technoscience involves acknowledging that cognitive encounters are only one way in which modern technoscience is lived.

In this chapter I have contended and demonstrated that the emergence of cultural studies of technoscience has been closely entwined with the history of late twentieth-century feminism. In virtually every strand of this field of critical scholarly research feminists have been in the vanguard, and feminist activism has inspired fresh insights and new orientations in approaches to modern technoscience. In this sense, the field of cultural studies of technoscience studies has not been a magical discovery. Nevertheless, it has been quietly transformative of contemporary understandings of technoscience and insistently optimistic about the possibilities of transforming both science and society.

Part I
Making heroes

3 Newton as national hero

In 1979, the historian of science Henry Guerlac reflected upon the progress of Newtonian studies and observed:

> Besides the technical study of Newton's achievements in mathematics, optics and dynamics, there is a phase of Newtonian scholarship which has attracted renewed interest and which we may call the 'influence', the 'reception', or the 'legacy' of Newton.
>
> (Guerlac 1979: 75)

This observation well describes the framework in which my initial research in this field was conceived. My own investigations originally focused on Newton's influence at the end of the eighteenth century in Britain, and case study 1 derives from that phase of my life when I was interested in the *impact* of Newton (pp. 28–32) and of his ideas. Hence, that analysis illustrates the kind of research Guerlac noted, which was prominent within historical and social studies of science during the 1970s and 1980s.

However, this chapter originated from my dissatisfaction with this mode of Newtonian studies. This unease was channelled into a presentation at a conference held in Oxford as part of the 1987 celebrations marking the tercentenary of the publication of Newton's *Principia Mathematica*. A revised version of that presentation (McNeil 1988) subsequently appeared in a collection entitled *Let Newton Be!* (Fauvel *et al.* 1988).

I had been invited to participate in the tercentenary conference because of my earlier research on the representation of Newton in eighteenth-century poetry (McNeil 1987). While I did draw on this work, I also used this occasion and the drafting of an article based on my conference presentation to reflect on my own movement within and away from what seemed to be the conventions of the historical and social studies of science in the 1970s and 1980s. Considering Guerlac's quotation in 1987 I articulated my unease with the mode of research and scholarship that he had identified. It struck me then that the approach he described needed to be taken further. I contended then that to understand what Newton – indeed, what any key figure in the history of science – means should involve investigating not only his writings and his influence, but also the other places and ways in which Newton is used as a symbol or lodges as a part of the popular imaginary. It was clear to me then, and, I would insist, is even more emphatically now, that this is not a simple matter of tracing influence or monitoring impact. Rather, it is about understanding the active creative processes whereby cultural meanings are generated about who Newton was, why he matters and what he comes to signify. In this sense, there is no single Newton. Beyond this, in my conference paper and subsequent article (McNeil 1988) I suggested that the celebration of the *Principia* could be the occasion for wider reflections on the particular significance of Newton in English national culture, the practices of science studies and the cultural processes through which science and scientific heroes in Western culture accrue meaning more generally.

Like my earlier article, this chapter revolves around a set of case studies focused on the mobilization of Newton as a key figure in English culture. Some of these derive from my own research and they range from detailed studies to commentaries which revolve around the identification of *potential* case studies. In addition, I have supplemented my own original work with a set of reviews of other case studies that have appeared since the publication of my earlier article (McNeil 1988).

The second dimension of the chapter is more reflective about the field of science studies: I highlight and assess the framing assumptions and methodological orientations which shape the various case studies presented here. In this respect, this chapter explores comparatively the parameters for investigating and interpreting the cultural crafting of Newton from the eighteenth century to the present. This involves tracing patterns and practices within the history and social studies of science and in the emerging field of cultural studies of science at the end of the twentieth and beginning of the twenty-first centuries. The case studies presented here and my tracking reflective appraisals are thus indicative of my personal trajectory, as well as movement within the field of science studies: from research within the history and social studies of science as they were generally practised from the 1960s into the 1980s, to forms of analysis oriented towards cultural studies of science.

Case study 1: capturing the English imagination – Newton in late eighteenth-century English poetry

The late eighteenth century is an interesting period for Newtonian studies, for there is a rich vein of Newtonian imagery in the poetry of the time – most notably, in that of Erasmus Darwin and William Blake. Darwin and Blake had much in common. Their social networks to some extent overlapped; they were both thought of as radicals and they explored within their poetry the world of natural philosophy. Indeed, Blake engraved some illustrations for one of Darwin's major poetic works, *The Botanic Garden* (1799). Nevertheless, their poetic visions were markedly different. Newton is a key figure within the poetry of each, but he is seen rather differently by these two poets. In each case, they mobilized their image of Newton in the construction of what Benedict Anderson (1983) calls 'imaginary communities', which underpinned the new sense of Englishness that they were striving to foster.

The work of Darwin and Blake came towards the end of a long debate about science within English poetry which had occupied much of the eighteenth century. Newton was, not surprisingly, a frequently encountered figure within this debate.[1] The particular features of Darwin's and Blake's images of Newton emerge within the context of this debate, which can be characterized as having four phases.

In the first phase, Newton's achievement was admired, but in a somewhat guarded or sceptical way, on grounds which are ultimately religious and to do with perspectives on the position of humankind in the cosmos. This phase is typified in the poetry of Henry Brooke and Alexander Pope. Brooke's 1730s assessment of Newton's importance had a clear perspective:

> For deep, indeed, the ETERNAL FOUNDER lies,
> And high above his work the MAKER flies;
> Yet infinite that work, beyond our soar;
> Beyond what Clarkes can prove, or Newtons can explore!
> ('Universal Beauty', 1735, Brooke 1778: 42, ll.319–22)

In *An Essay on Man* (1733), Pope expressed similar reservations about Newton's accomplishments:

> Superior beings, when of late they saw
> A mortal Man unfold all Nature's law,
> Admired such wisdom in an earthly shape,
> And shew'd a Newton as would shew an Ape.
> (Pope 1753: 34, ll.31–4)

These poets saw Newton as embodying the strengths, but also the limitations, of the human comprehension of the natural world.

In the second phase, it was the imagery which Newton had unleashed that captured the imagination of poets. His *Opticks* (1704), in particular, was a major influence on the work of many poets, for whom Newton had intensified the human experience of the natural world. In the middle of describing the rainbow, James Thomson addressed Newton directly:

> Here, awful Newton, the dissolving clouds
> Form, fronting on the sun, thy showery prism;
> And to the sage-instructed eye unfold
> The various twine of light, by thee disclosed
> From the white mingling maze.
> ('To the memory of Sir Isaac Newton', 1728,
> Thomson 1908: 11, ll.208–12)

Thomson wrote this poem in 1728, the year after Newton's death. This phase of drawing upon Newtonian imagery was still vigorous in the 1740s, when Mark Akenside (1744: Book II, ll.100–20, pp. 51–2) conveyed through his poetry his expectation that the human experience of rainbows would be qualitatively improved through Newton's theories.

A rather different response to Newton can be seen in the poetry of Christopher Smart and William Blake. Both of these poets were dissatisfied with the world-picture implied by the works and reputation of Newton and his friend the philosopher John Locke. Smart and Blake regarded the influence of Newton and Locke as negative, because their views obscured crucial questions about the relationship of knowledge, truth and beauty. Indeed, these poetic critics seem to verge on a kind of materialism which left no space in the universe for God. Such a reaction against the earlier admiration for Newtonian views was part of what lay behind Blake's protest in *A Vision of the Last Judgement* (1810):

> The Last Judgement is an Overwhelming of Bad Art & Science. Mental Things are alone Real; what is call'd Corporeal, Nobody Knows of its Dwelling Place: it is in Fallacy, & its Existence an Imposture. Where is the Existence Out of Mind or Thought?
>
> (Blake 1972: 617)

In poetry which appeared in the last decade of the eighteenth century a fourth phase emerges, still inspired by Newton. Poets such as Richard Payne Knight and Erasmus Darwin were optimistic, as Thomson and Akenside had been, about the promise of contemporary natural philosophy. Yet they shared with Smart and Blake doubts about whether the picture of a world governed by Newtonian natural laws actually made religious sense. Knight, for example, pondered:

> Whether primordial motion sprang to life
> From the wild war of elemental strife;
> In central chains, the mass inert confined
> And sublimated matter into mind? –
> Or, whether one great all-pervading soul
> Moves in each part, and animates the whole;
> Unnumber'd worlds to one great centre draws;
> And governs all by pre-established laws?
>
> (Knight 1796: 3, ll.1–8)

In fact, Knight concluded that he could not tackle these issues, beckoning his readers:

> Let us less visionary themes pursue,
> And try to show what mortal eyes may view;
>
> (Knight 1796: 3, ll.15–16)

Darwin was more forthright in this respect. For him, the only way of justifying claims to have acquired true knowledge about nature was to frame these in an account of how the natural world and the human mind shared a common origin. By extending the ambition of Newtonian natural philosophy, Darwin sought to bring all social, intellectual and moral developments under the umbrella of the operations of nature:

> By firm immutable laws
> Impress'd on Nature by the GREAT FIRST CAUSE.
> Say, Muse! how rose from elemental strife
> Organic forms, and kindled into life;
> How Love and Sympathy with potent charm
> Warm the cold heart, the lifted hand disarm;
> Allure with pleasures, and alarm with pains,
> And bind Society in golden chains.
> (Darwin 1803: Canto I, p. 1, ll.1–8)

Later in this posthumously published poetic work, *The Temple of Nature* (1803), Newton appears in Darwin's pantheon of heroes, along with those stalwarts of the Industrial Revolution – Newcomen, Arkwright, Priestley, Savery, Wedgwood and Boulton:

> ... NEWTON's eye sublime
> Mark'd the bright periods of revolving time;
> Explored in Nature's scenes the effect and cause,
> And, charm'd, unravell'd all her latent laws.
> (Darwin 1803: Canto IV, p. 148, ll.233–6)

Darwin constructed an 'imaginary community' (P. Anderson 1969) of scientists and industrialists, whose achievements he regarded as the culmination of the development of the natural world. Here were his true English heroes and his poetry celebrated their accomplishments.

But there was a further dimension to Darwin's poetic projects which brought him back to Newton. Darwin's poetic images of the natural and social worlds were designed, as he put it, in the Advertisement of *The Botanic Garden* (Darwin 1799) 'to inlist the Imagination under the banner of Science'. He explained there that the imagination was to be the vehicle mediating sensual poetic images and science, which he called 'the ratiocination of philosophy'. But the imagination was also, for Darwin, something that was very visually oriented, as his poetic practice and theory demonstrated. As he explained, 'the Poet writes principally to the eye' (Darwin 1803: Advertisement). Indeed, Darwin's conception of knowledge acquisition was founded on a visual model, and he explored various ideas concerning the physiological dimensions of vision – including speculation about 'luminous music', drawing on Newton's theory of the colour spectrum.

Although his attitude to Newton was very different, William Blake was also concerned with Newton's status as the major figure in contemporary natural philosophy. The core of the appeal of the Newtonian cosmology, in Blake's estimation, was its 'picture-language quality' (Ault 1974: 50), which drew on both the imagination and reason. While Darwin viewed this as a positive feature of Newtonian natural philosophy which he wished to emulate, Blake saw it as a real threat to the human imagination. He endeavoured to extract the imagination from its Newtonian constraints and was set firmly against Darwin's aim of its mediating between science and poetic images. Blake's intention was to subvert *images of vision* by transforming them into *visionary images*:

> The nature of infinity is this: That every thing has its
> Own Vortex, and when once a traveller thro' Eternity
> Has pass'd that Vortex, he percieves [sic] it roll backward behind
> His path, into a globe itself unfolding like a sun;
> Or like a moon, or like a universe of starry majesty,
> While he keeps onwards in his wondrous journey on the earth,
> Or like a human form, a friend (with) whom he liv'd
> benevolent.
> ('Milton', written and etched 1804–8 in Blake 1972: 497, ll.21–7)

So Darwin and Blake took opposing stands on the imagination, reason, vision and, indeed, on industrialization. For Darwin, Newton was a hero, and readers were to be drawn into the world of science through poetry, in particular through appeals to the visual sense and thereby through an instrumentalist use of the imagination. For Blake, Newton was equally important, something of an anti-hero – linked in his own mythology to the figure of 'Urizen', the tyrant who threatened to dominate and destroy the universe. While the achievements of Newton, Watt, Savery and Priestley represented progress for Darwin, for Blake they were identified with the impoverishment of the imagination. 'Urizen' symbolized his fears about the dangers of Darwin's goal of enlisting 'Imagination under the banner of Science'. Blake was determined that this and other complementary aims should not be realized on Albion's fair shores.

Case study 1: some reflections

When I began to examine the issues just described, I was operating, albeit unselfconsciously, within the model described by Guerlac (1979). I too was in pursuit of Newton's influence – in this case, on Darwin and Blake. This framework now seems unsatisfactory. I would argue for a broader perspective than pursuing one particular hare which wanders around late eighteenth-century Newtonian studies – namely the question of who has or had the *right* version of

Newton. Instead, I would prefer to see Darwin's and Blake's projects as embodying constructions or reconstructions of Newton. Each provided the basis for substantive calls for a reorientation in English culture. In Darwin's case, this involved adaptation to the goals of the new innovative bourgeoisie of provincial England. Blake, by contrast, was seeking an animated, anti-industrial, imaginative culture to pose as an alternative at the start of the nineteenth century. The visions of Darwin and Blake were counterpoised: Newton as hero or Newton as anti-hero; a new England, regenerated 'under the banner of Science', or confined by the 'Loom of Locke' and 'the Water-wheels of Newton'. Each built his aspirations for Albion on different, but complementary, images of Newton.

Case study 2: Newton and modern English national culture – left assessments in 1960s and 1970s Britain

My first case study of Newton as a figure in English national culture was focused on the eighteenth and early nineteenth centuries. In many respects, it felt riskier to tackle more recent conjurings of him, since this seemed to carry me further from the domains Guerlac had specified. Nevertheless, in the mid-1980s I became aware of Newton's significance within at least one explicitly political debate about English culture towards the end of the twentieth century which was intriguing. Tracing this debate yielded another, more contemporary case study in the making of Newton.

In the late 1960s some English intellectuals of the left, in an exercise remarkably similar in substance to that of Erasmus Darwin and William Blake traced previously, although different in form, reviewed the 'components of the national culture' (P. Anderson 1969). Some reached dismayingly negative conclusions. Perry Anderson (1969: 215), for example, concluded that England had a culture that was 'mediocre and inert', although it is perhaps significant that he deliberately excluded the natural sciences from his survey. This and other negative appraisals inspired Edward Thompson to a spirited defence of his national culture. In an article with the marvellously ambiguous title 'The peculiarities of the English' (E. P. Thompson [1965] 1978), Thompson expounded on the richness of the maligned national culture. In contrast to Anderson, Thompson's main reference point in his rehabilitation was English natural science. The hero of his version of the national culture was Charles Darwin, and standing in the wings to provide trusty, predictable support were those key figures in the English natural philosophy tradition – Francis Bacon and Isaac Newton. Here was the core of the 'peculiar' English intellectual tradition, as Thompson identified it. Natural philosophy was highlighted as the English intellectual discipline *par excellence*.

Returning to this trail of debate leads to my first suggestion about the more contemporary significance of Newton within English culture. There is evidently something about this natural philosophy tradition, represented by the triumvirate of Bacon, Newton and Darwin, which has come to represent English culture to a wide public.

Case study 3: Newton and the apple – a key myth

Thompson's explicitly politically motivated genealogical investigation highlights Newton's importance in English intellectual life. But there are other more popular Newtonian symbolic trails in English cultural life which also merit explication. In an intriguing set of essays examining some key contemporary myths, first published in 1957, the French semiologist Roland Barthes turned his attention to Albert Einstein. As his essay title 'The brain of Einstein' suggests, Barthes was intrigued by what seemed to him to be an obsession with this particular part of Einstein's anatomy. As he saw it, Einstein's brain was the avenue through which this scientist was most frequently encountered in popular culture. This image itself was contradictory, he maintained: 'Einstein's brain is a mythical object: paradoxically, the greatest intelligence of all provides an image of the most up-to-date machine' (Barthes 1973: 68).

But Barthes pursued the mythology surrounding Einstein yet further, considering that '[t]hrough the mythology of Einstein, the world blissfully regained the image of knowledge reduced to a formula' (Barthes 1973: 69). Even more than this, in Barthes's estimation Einstein was a crucial figure in popular culture because he

> fulfils all the conditions of a myth, which could not care less about contradictions so long as it established a euphoric security: at once magician and machine, eternal researcher and unfulfilled discoverer, unleashing the best and the worst, brain and conscience. Einstein embodies the most contradictory dreams, and mythically reconciles the infinite power of man over nature with the 'fatality' of the sacrosanct, which man cannot yet do without.
>
> (Barthes 1973: 70)

Barthes provides a cue relevant in considering the mythologizing around Newton. The most popular mythical representation of Newton revolves around the story of the apple. According to this story, Newton's understanding of gravity and, indeed, his subsequent formulation of the laws of motion of bodies were inspired by the observation of the simple phenomenon of an apple falling as he lay under a tree in his garden. Or, in the bold claim of the magazine *Garden Answers* (September 1982: 29), an apple falling on his head 'caused him to discover gravity'. Whatever the sources and reality of this tale, it does have extraordinary popular purchase. Indeed, the British Royal Mail presumed complete public familiarity with the image of the apple as denoting Newton when they chose it as an appropriate emblem for the stamp which marked the *Principia* tercentenary in 1987.

It is precisely the popular appeal of this myth which demands closer scrutiny. What is it that makes it so appealing, such a recurring reference? We might return to Barthes' description of Einstein as a mythical figure who 'reconciles the infinite power of man over nature with the "Fatality" of the sacrosanct, which man cannot yet do without'. The myth of Newton's apple seems to fulfil similar needs. Is there not something marvellously reassuring about this tale? Nature quite literally knocks us on the head with the required knowledge! The story bespeaks an amazing integration between humanity and the natural world. What better endorsement could there be for the potential of the 'gentleman scholar'? In addition, this is fundamentally a pastoral image, as are so many other cherished images of Englishness, including the paintings of Constable: it is, thereby, associated with that touchstone of Englishness – the countryside.

This simple and often repeated tale resonates because it is so reassuring. Amidst the insecurities generated by nuclear and nanotechnology and the big moral questions posed by genomic and reproductive technologies, the story of Newton's apple offers a more peaceful image of the generation of scientific knowledge – one which is fundamentally without conflict. Moreover, it is also reassuring about Britishness in general, and about British knowledge of the natural world in particular. It reanimates British pastoralism, whilst insisting that, whatever disclaimers are made about the state of British science and technology today, it has been, and perhaps still is, powerful.

Mapping further case studies: Newton in contemporary popular British culture

Newton's apple deserves attention as one of the most iconic images which circulates with this English scientific hero. Nonetheless, as I suggested previously (McNeil 1988), there are many other concrete and empirically rooted ways of pursuing the contemporary significance of Newton. I proposed then and would now reiterate that one place to begin might be with a review of the various potential sites of encounter with Newton which exist in contemporary British popular culture. This carries us outside the academic context, to the ways that 'the woman or the man in the street' may meet this scientific hero. This avenue of investigation is extensive and multidimensional. Nevertheless, it seems a potentially fruitful area of investigation for those concerned with the role of science in culture, with national heroes and/or with popular culture.

For this reason, it might be worthwhile to sketch some of the parameters of such research on Newton. Inhabitants of twenty-first-century Britain might encounter Newton in any or all of the following ways:

- as noted previously, through the postage stamps issued in 1987 to mark the 300th anniversary of the publication of the *Principia*;
- in school textbooks, cited as a major figure in the history of science;
- as a figure represented in renowned statues in some key national settings, such as Westminster Abbey or Trinity College, Cambridge;
- as a person who was represented on the old pound-sterling note, thereby joining other notables (such as Shakespeare or the Duke of Wellington) whose images have adorned British currency.

Even this cursory and limited listing is revealing about Newton's role in contemporary British culture. Each of the instances listed above involves a key institution of British life: textbooks within the educational system, statues at the national church of the state religion (Westminster Abbey) and one of the oldest established English universities (Cambridge), currency and national financial institutions, and the postal service, as one of the oldest and most important components of the national communications system. Indeed, currency and postage stamps are amongst the most tangible, ubiquitous and immediate international signifiers of national identity. Westminster Abbey and Cambridge University have double valence, both as seats of major establishment institutions (religion and higher education) and as tourist sites and venues for homage to the relics and paraphernalia of Englishness. Textbooks function, perhaps, more as internal carriers of national identity. Linguists, historians and cultural analysts have acknowledged their importance in this respect, seeing them as repositories of crucial images of nations.

Despite the variety and dispersal of locations in which Newton might be encountered, most of these sites do have particular national resonance in British culture. In this sense, Newton is an intensely mobilized figure in what Michael Billig has termed 'banal nationalism'. Billig explains that

> in the established nations [and he includes the United Kingdom in his list of these], there is a continual 'flagging', or reminding, of nationhood ... [which] provides a continual background for their political discourses, for cultural products. ... In so many little ways, the citizenry are daily reminded of their national place in the world of nations. However, this reminding is so familiar, so continual, that it is not consciously registered as reminding.
>
> (Billig 1995: 8; my interjections)

These representations within key national institutions are indicative of Newton's status as a powerful signifier of Britishness. However, as Billig suggests, this takes the form of encounters which are mundane and not necessarily consciously registered as nationalistic.

Newton in British culture: from the late eighteenth century to the present

In the preceding sections I have explored some dimensions of Newton's role within British culture in different periods. It is appropriate now to draw out some features of my case studies. In each of the twentieth- or twenty-first-century instances sketched earlier, Newton clearly signifies *past* achievements identified with British culture. This brings us back to case study 1. There we saw Darwin and Blake thrusting Newton into the centre of their projects for major reorientations of English culture. For Darwin, Newton was at the heart of his ambition to bring his contemporaries to the world of science and industry, and of his intent to make contemporary institutions take account of science and industry (and scientists and industrialists). Blake, in contrast, saw Newton as a major obstacle to a more visionary and imaginative, less materialistic culture. Both poets were inspired by Newton to beckon their contemporaries towards a new kind of Englishness. Theirs was a vision of what Albion might be – a *future* vision.

Reading the poetry of Darwin or Blake, their contemporaries could link Newton either positively or negatively with a call for change, for a reorientation in English culture. In contrast, more contemporary encounters with Newton are not likely to be about change. They are more likely about what British culture was, and is to remain. Newton inspired Darwin and Blake, and by extension some of their readers, with a sense of what English culture *could* be. In the late twentieth and early twenty-first centuries, the popular encounter is with a Newton who represents British culture of the past. He is generally associated with a Britishness which is to be preserved, and not changed. He is met as part of what Robert Hewison designates as 'heritage culture' (Hewison 1986: 308; 1987), or as 'the historical and sacrosanct identity' (Wright 1985: 2) of the British nation. Newton's contemporary role would seem to be part of the more general pattern that Patrick Wright has described, whereby 'the past has been secured as a cultural presence in modern Britain' (Wright 1985: 3).

In both periods, Newton comes to represent some aspects of British science. In Darwin's and Blake's era, he represented the potential of natural philosophy: positively for Darwin, negatively for Blake. By contrast, in the late twentieth and early twenty-first centuries Newton appears as a figure of reassurance rather than promise: identified with the past rather than the future of British science.

It is also interesting to contrast the settings in which Newton appears. In case study 1, I examined some poetic explorations of Newtonianism. This literary mode was pre-eminent in encouraging public understanding of Newton's natural philosophy in the eighteenth century. In the twentieth and twenty-first centuries, the lay encounter with Newton is less likely to occur via literature, even though there have been recent novels in which Newton is a key figure (Beaven 1994; Banville 1999).[2] In fact, it could be claimed that Newton is now much more integrated into popular culture at the everyday level than he was in late eighteenth- or early nineteenth-century Britain.

There are two further observations which emerge from juxtaposing the case studies considered earlier. One is that Newton has been a pivotal figure within British culture for over two hundred years. His significance for Darwin and Blake was that they regarded him as a lynchpin in their efforts to reorient English culture – he was crucial to their vision of the future of England. Today, Newton is more likely to be taken as a signifier of the past – ensconced as part of British heritage, and as part of a British culture which is to be preserved rather than changed. Finally, his potency as a signifier of Britishness has grown since the late eighteenth century. Represented in major institutions of the national establishment, and linked to pastoral traditions, Newton, like Shakespeare, has come to be a powerful symbol of Britishness.

The figure of the scientist: history of science and cultural studies of science

The history of science is a discipline which, until fairly recently, was, to a considerable degree, organized around key figures, including Newton. This biographical focus still casts its shadow over research and teaching undertaken within the discipline. Some conferences and books (including one mentioned previously – Fauvel *et al.* 1988) have as their organizational core Newton's persona, writings, interests and activities. There is a certain intellectual quest which motivates much of the input into such productions: identifying the *real* Newton. What were his prime interests? What are his theories and writings really about? Are there elements of his activities and enterprises which have not been adequately acknowledged or analysed?

These are all legitimate and important questions in the history of science. Nevertheless, our historical understanding of science in culture, and of Newton as a particular figure within it, has sometimes been enriched when we are willing to let go of the pursuit of the *real* Newton. In short, there is much to be gained by, in some instances, shifting focus away from Newton himself, to versions of his ideas and to the constructions and reconstructions of his significance. Hence, we can learn more from investigating how Darwin and Blake used Newton in their reorientations of English culture than from evaluating how *true* their different versions were to *the real* Newton. Likewise, contemporary myths about Newton, including the story of his enlightenment through his observation of a falling apple, can be investigated to yield rich information about Newton's changing significance in British culture.

A second major feature of the history of science has been its preoccupation with key texts. Gyorgy Markus (1987) has argued that the natural sciences themselves have for a long time been structured around key texts which *belong* in a strong sense to their author, like an intellectual patent. The orientation of historical studies of science around major scientific texts has, in this respect, paralleled the natural sciences themselves. I do not undermine the importance of this work in suggesting that interpreting Newton as a cultural hero can also involve studies of *other* texts and of *other* cultural sites. The contemporary meaning of Newton is produced through encounters with stamps and with statues of him, as well as with the *Principia* itself. As this indicates, cultural studies perspectives challenge the narrow text orientation of some history of science.

The final point I wish to draw from my analysis takes me back to the well-trodden terrain of preceding debates within the discipline of history of science about what kind of factors influence the development of science. It is significant that Newtonian studies have been a battleground for such controversies. For a period there was a lively debate within the history of science about whether Newtonian mechanics was a product of a particular state in the development of capitalism or the product of internal development within the study of the natural world. The fervour animating the clashes between *externalists* and *internalists* within the history of science – as proponents of these two positions were called from the 1960s into the 1980s – made it difficult to note that there was something of a convergence of focus between these two approaches (Shapin 1992). For, whether they drew on internal or external factors, the object in both cases was to explain developments *within* science. In short, they shared a common object of study. Much of the history of science has drawn its cues, whether internalist or externalist, from science itself in this focused sense. The logic of this orientation has gone something like this: to understand science we must begin (at least) with the scientific community, the scientist or the scientific text, and possibly then move outwards in other directions.

I am suggesting that cultural studies of science undertakes rather different projects. Such work is premised on the notion that the meanings of science are not all rooted in the scientific community itself. This is not a straightforward issue of 'influence' as Guerlac and others have proposed, for it can involve the generation of distinctive meanings about science in a variety of sites. A cultural studies perspective acknowledges that the contemporary meaning of science is not restricted to the domain of its formally designated practitioners, but rather registers that it is constructed across a range of disparate locations and material forms, including, as suggested previously, tourist sites, postage stamps and so on. Newton's figure in these locations is just as much a locus of meaning as that reconstructed by historians poring over his primary texts. Indeed, this chapter has only hinted at the range of research projects which might be undertaken to provide a fuller picture of Newton's significance in popular English culture. Such a strategy by no means denigrates the stature of Newton. Indeed, what I am proposing is a fuller and more complex investigation of why and how Newton has been such a crucial national hero.

Case study 4: Richard Yeo's study of British scientific culture in the early nineteenth century

My proposal derives from the hope that the scholarly fixation with the identification of *the real* Newton would give way to the investigation of the makings of Newton in a range of specific historical contexts. As it happened, and unbeknownst to me when I wrote my article (McNeil 1988), a project investigating the making and remaking of Newton in Britain between 1760 and 1860 had already been undertaken by Richard Yeo and this was presented in an article published in *Science in Context* in autumn 1988 (Yeo 1988). This study centred on the related, but distinct, versions of Newton generated by key figures involved in creating and defending a space for science in British culture in the early nineteenth century – David Brewster, William Whewell and Augustus De Morgan. Yeo traced the changing profile of Newton during this period, showing how developing conceptions of genius and scientific method refashioned this figure. The unearthing of evidence of a range of new materials pertaining to the celebrated natural philosopher's diverse engagements in alchemy, sectarian religious disputes including Tractarianism, and family scandals relating to his niece were important elements in Yeo's research. This research informed his interpretations of emerging early nineteenth-century notions of the romantic secular genius, and medical and psychological theories of insanity and genius that were crucial to the recastings of Newton.

In effect, Yeo (1988) offered a very striking case study of the kind I had proposed in my article published the same year (McNeil 1988). He demonstrated concretely the cultural crafting processes and contestations that were crucial in the making of Newton as a scientific hero over a one-hundred-year period. The implication of this project would seem to be that change in versions of Newton would not be limited to the period Yeo had studied, but he does not speculate beyond his own work or reflect on the broader significance of his research. His conclusion constitutes a specific reflection about the time period he investigated, but, strikingly, this is offered in the trope of secular progress. Referring to the 'changing of images of Newton' during this period, Yeo concludes that '[t]he result was the loss of a number of assumptions by which Newton's science was tied to a theological and moral framework and, arguably, the emergence of a more valid historical image of Newton' (Yeo 1988: 279).

Patricia Fara's case studies of the making of Newton

My own 1980s research was more attuned to Patrica Fara's more recent contention that 'no "true" representation of Newton exists' (Fara 2003: xvi). Fara is the scholar who has most systematically pursued the documentation and analysis of the various and diverse makings of Newton. Two articles – one on 'images of Newton in eighteenth-century England' and one on key memorial sites celebrating Newton (Fara 2000a, 2000b) preceded and anticipated her major publication, *Newton: the making of genius* (Fara [2002] 2003). Combining a thematic and a chronological presentation, Fara analyses a panoply of representations of Newton, tracing the construction of him as a genius and, in Britain, as a national hero from the period of his natural philosophical work in the seventeenth and eighteenth centuries to the start of the twenty-first century. She explains: 'Rather than searching for more facts about Newton himself, this book explores how he became celebrated as a national hero and a scientific genius – a secular saint for our modern society' (Fara 2003: 3). The result is a textured tapestry which displays the complexity of this historical figure and into which are woven countless cameos about his fabrication, highlighting key locations, agents, materials and cultural processes. Fara realized the kind of analysis that I could only gesture towards in my earlier article. This can be illustrated by reviewing two common foci that link my 1988 article and Fara's subsequent volume.

As case study 3 indicates, my article (McNeil 1988) signalled and reflected my own fascination with the power and significance of the apple as a widely recognized symbol of Newton, exemplified in its use on a stamp issued by the British Royal Mail in 1987 to mark the tercentenary of the publication of the *Principia*. Fara also registers the currency and purchase of this image and of the related story, noting that '[m]ost people know very little about Newton's physics, but they do know that he watched an apple fall from a tree' (Fara 2003: 192). She traces the origin of this story and the history of its take-up, explaining that '[a]lthough Newton himself originated the apple story, it was scarcely known until the early nineteenth century' (Fara 2003: 194). In fact, Fara attributes the first articulation of the story for a popular audience to Benjamin Martin in 1764 (Fara 2003: 216). She situates this tale within a constellation of mythical images and constructs that have come to be associated with Newton since his lifetime. In addition to pinpointing the origins and pattern of take-up, Fara identifies some of the connotations of this Newton myth in his period which did not linger beyond the early eighteenth century (Fara 2003: 198). She also scrutinizes specific episodes in the adaptation of the myth to serve different interests and purposes, particularly in the nineteenth century.

In general, then, Fara's research situates my observations and suggestions about the resonance of the apple image in late 1980s Britain in a much broader historical picture. It also elaborates on the riffs in the rich repertoire of associations which have accrued around this image in modern Britain, underscoring its importance in providing an 'Arcadic' (Fara 2003: 5, 197) model of the acquisition of scientific knowledge. Indeed, in a book deriving from a recent BBC radio series on 'Great Scientists' Melyvn Bragg marvelled at the lingering appeal of the story of Newton's apple, observing: 'The idea of thinking about gravity when an apple fell has everything. It is so simple it seems like something out of an ancient myth and it is something with which every child can instantly identify' (Bragg 1998: 80).

Both Yeo (1988) and Fara (2003) trace how the image of Newton became a resource in the ongoing negotiations regarding the significance of science within British culture, which I tackle through the case studies I outlined earlier in this chapter. Case study 1 analyses some key representations of Newton in eighteenth-century English poetry; case study 2 revolves around Newton's recruitment in twentieth-century political evaluations of English culture. I also sketched some starting points for would-be research projects relating to the contours of Newton's profile in contemporary Britain. In relation to my earlier article (McNeil 1988) and Fara's (2003) subsequent book, Yeo's (1988) project might now be regarded as a crucial and complementary case study. While case study 1 considered how poets used the figure of Newton to expound and explore expectations and fears about the growing powers of natural philosophy in eighteenth-century English culture, Yeo (1988) focuses on a group of academics who played a more direct role in creating space for science and scientists in early nineteenth-century Britain – a very early, but crucial, period in the designation of this field and this role.[3]

Twenty-first-century reflections on doing cultural studies of science

The juxtaposition of my earlier article (McNeil 1988) with Yeo's (1988) and Fara's (2000a, 2000b, 2003) studies brings into relief two theoretical and methodological devices deployed in the structuring of my analyses. The use of the term case study is a way of registering self-consciousness about interpretative processes and a concern about investigating detailed cultural processes whilst giving attention to broad cultural patterns. For me, this is clearly a gesture in moving away from naturalistic discourses and modes of writing characteristic of traditional historical research which attempts 'to tell it like it is'.[4] In a related way, the employment of Benedict Anderson's concept of 'imaginary community' involves the importation of a theoretical pivot for the study in the hope that it would both enrich and denaturalize the historic account I was assembling. It now seems to me that these were important methodological moves (about which I was by no means fully conscious in the late 1980s), which helped to forge versions of doing cultural studies of science rather than pursuing the more conventional modes of history and social studies of science of that period.

Of course, disciplines are not static, and in the late 1980s it was easier to make the history of science a bit of a straw man by critically highlighting its lacunae than it is today. Moreover, interdisciplinarity has generally found more favour and infiltrated into a wide range of disciplinary practices. So, for example, cultural studies perspectives have also percolated into the history of science. Increased explicitness about theoretical and methodological dimensions of research within this discipline and the self-designation of 'cultural historians' signal these transformations (Chartier 1988).

A recent exemplification of these developments in history of science research can be found in a publication in which Ludmilla Jordanova also uses the concept of 'imagined communities' to structure her exploration of 'the ways in which science and nationhood became intertwined between the eighteenth and twentieth centuries' (Jordanova 1998: 195). She highlights, in particular, the development of biography and portraiture as 'central genres not only for the construction of medical and scientific identity, but also ... for the working-out and management of relationships between nationhood and the practice of science and medicine' (Jordanova 1998: 193) in Britain in the eighteenth and early nineteenth centuries. Jordanova underscores processes of identification as the analytical fulcrum for her investigation of the emergence of national imaginary scientific communities. While she ranges widely in tracking the making of such a community in Britain, particularly between the eighteenth century and early nineteenth century, she hones in on practitioners of science and medicine as the key agents in this process.[5] There is something of a tension, then, between, on the one hand, her attribution of distributed agency in the constitution of the imagined British scientific community implied by her detailed analysis of specific portraiture and biographical writing and, on the other hand, this specification of scientists as *the* agents in this cultural bonding.

In venturing into the field of Newtonian studies in 1987 I was aware that I was entering territory shaped by the concerns and structures of the established discipline of history of science of that period. It struck me then as ironic that a discipline so preoccupied with a few key heroic figures should not be interested in the related investments in heroes of science that was occurring in other cultural locations. On the other hand, it was precisely the academic pursuit of the *real* Newton which served to trivialize and delegitimize other cultural modes of pursuing or, as I (and Fara) would insist, *making* Newton. Although more traditional Newtonian scholarship continues apace, history and social studies of science have been reconfigured, placing more emphasis on the studies of social patterns and cultures of technoscientific practice (see, for example, Shapin and Schaffer 1985). While this has encouraged more awareness of the multiple and dispersed makings of meanings in science and dissipated the focus on heroic individuals, there continues to be resistance to engagement with popular culture in some quarters.

Nonetheless, from within the history of science community a few voices have been raised, questioning this lacuna in historical research on science, making the case for giving more attention to those not formally designated as actors on the scientific stage. Most notably, Roger Cooter and Stephen Pumphrey lament that 'surprisingly little has been written on science generally in popular culture, past or present' (Cooter and Pumphrey 1994: 237). Cooter and Pumphrey provide an impressive review of would-be resources within and around the fields of history and social studies of science in an attempt to explain this 'absence' (Cooter and Pumphrey 1994: 239). Their survey is a clarifying resource for those interested in trying to investigate science in popular culture and it is a valuable repository in configuring the challenges of doing cultural studies of science. What can be carried into the framing of the case studies presented in this chapter is Cooter and Pumphrey's considered evaluation that developments in the history of science, in the sub-discipline of public understanding of science and in the new sociology of science into the mid-1990s were all actively blocking research into science in popular culture, rendering it, 'in effect, ... an intellectually "inauthentic" sphere' (Cooter and Pumphrey 1994: 246). Although the general assessment of the state of these research fields and the exploration of related theoretical and linguistic difficulties offered by these historians of science is not optimistic, they do highlight instances of exemplary research and promising theoretical developments. They share the concern which prompted my investigation of Newton as a national hero, that the history and social studies of science should 'be responsive to a greater plurality of the sites for the making and reproduction of scientific knowledge' (Cooter and Pumphrey 1994: 254). It is this concern which has drawn researchers to find modes of investigation appropriate in undertaking cultural studies of science.

4 Making twentieth-century scientific heroes

> By virtue of their ascribed feminine characteristics, women hardly fit the category of 'scientist' at all.
>
> (Van Dijck 1998: 24)
>
> Science seldom proceeds in the straightforward logical manner imagined by outsiders. Instead, its steps forward (and sometimes backward) are often very human events in which personalities and cultural traditions play major roles.
>
> (Watson 1969: ix)

Chapter 3 considered different versions of Newton as a British national scientific hero which emerged over several centuries. This chapter, in contrast, explores the making of scientific heroes in the last third of the twentieth century. As noted in Chapter 3, Newton's achievements have been marked and celebrated in a diffuse range of British cultural sites. Thus, in terms of cultural forms the scope of my analysis was broad, ranging across a variety of sites of Newton's figuration: from eighteenth-century poetry to twentieth-century postage stamps. This chapter is focused on a very specific cultural form – popular science biographies that appeared in the English-speaking world between 1968 and 1983. It explores the making of new scientific heroes of the late twentieth century in and through this distinctive cultural form.

The chapter begins with brief introductions to the four books that are my central concern. I start with the most popular scientific memoir of the twentieth century, James Watson's *The Double Helix* ([1968] 1969), and I then introduce three widely circulated biographies of women scientists of the twentieth century: Ann Sayre's *Rosalind Franklin & DNA* (1975), June Goodfield's *An Imagined World: a story of scientific discovery* ([1981] 1982) and Evelyn Fox Keller's *A Feeling for the Organism: the life and work of Barbara McClintock* (1983).

The books

The four books that are my central concern are as follows:

- James Watson, *The Double Helix* ([1968] 1969): this text constituted Watson's personal account of the research associated with the discovery of the double-helix structure of DNA. The book recounts the collaboration between

Watson (a young American researcher who had recently earned his PhD) and Francis Crick (an older, English PhD student), working in the Cavendish Laboratory at Cambridge University between 1951 and 1953, which led to the construction of a model of DNA and the publication of a key paper on this topic in *Nature* (April 1953). This research resulted in the award of a Nobel Prize to Watson and Crick, together with Maurice Wilkins (King's College London) in 1962.

- Ann Sayre, *Rosalind Franklin & DNA* (1975): this biography of physical chemist and crystallographer Rosalind Franklin (1921–58) was written by the scientist's friend, who was also the wife of one of her colleagues. The book attempts to put the record straight about Franklin's role in the discovery of the double-helix structure of DNA. It contests the account of this discovery provided by James Watson in *The Double Helix* and the negative portrayal of Franklin offered in that text. As just noted, Watson, Crick and Franklin, along with their former colleague at King's College, London, Maurice Wilkins, were awarded a Nobel Prize in 1962 for their work on the structure of DNA. Rosalind Franklin died of cancer in 1958 at the age of 37.
- June Goodfield, *An Imagined World: a story of scientific discovery* ([1981] 1982): Goodfield, an established historian of science and scientific journalist, documents the work of Dr Anna Brito (the pseudonym used in the book) (b.1942), a Portuguese immunologist whose research carries her to posts in London, Glasgow and New York. The account is based on Goodfield's observation and extensive communications (interviews, letters, phone calls, tape recordings) between the author and Anna from 1975 to 1980. It focuses particularly on Anna and her international research team based in their laboratory in New York as they investigate the role of iron in the functioning of the immunological system, especially in dealing with cancer.
- Evelyn Fox Keller, *A Feeling for the Organism: the life and work of Barbara McClintock* (1983): Keller, a former physicist, turned historian of science and feminist theorist, wrote this biography of the US geneticist Barbara McClintock (1902–92). McClintock was awarded the Nobel Prize for Medicine and Physiology shortly after this book appeared (1983) for her work on gene transposition, which showed that genes 'jump', that their behaviour is random and that they can move between cells.

While Watson produced the most popular scientific memoir of the twentieth century, these other books were the most widely circulated accounts of women's lives as working scientists available in the English-speaking world in the 1970s and 1980s. Watson's narrative haunts the other texts. This is most obvious in the case of Sayre's book, which is a direct rebuttal of *The Double Helix*, but is apparent in the other books as well. There are striking connecting threads and intertextual references linking these three biographies of female scientists. June Goodfield's Anna reads and responds to Sayre's book on Franklin: her response to this text informs her perspectives on difficulties women experience in pursuing scientific careers. Both Goodfield and Keller use the notion of 'winners' and 'losers' in science (pertaining to scientific discovery), which was the leitmotif of James Watson's *The Double Helix* and was adapted more critically to frame Sayre's story of Franklin.[1] Keller (1981), who reviewed Goodfield's book, quotes it directly in the conclusion of McClintock's biography, linking the Nobel Prize winner's methodological approach – her 'feeling for the organism' – with Anna's orientation towards research summarized in the younger scientist's observation that '[i]f you want to really understand about a tumor, you've got to *be* a tumor' (Keller 1983: 207; Goodfield 1982: 226).

The Double Helix: rewriting scientific heroism

As I have outlined, *The Double Helix*, which was first published in 1968, was James Watson's personal memoir of the research associated with the discovery of the double-helix structure of DNA. The pre-publication controversy surrounding this book indicated that there was a lot at stake in putting this personal account into the public domain. Watson had circulated drafts of the text to a number of colleagues, and his two co-holders of the Nobel Prize for the discovery of the double-helix structure of DNA – Francis Crick and Maurice Wilkins – objected to its publication. In an unprecedented decision, the Board of Harvard University Press decided to halt publication (Stent 1980a; Yoxen 1985). Watson published with Atheneum Press instead and the book became a bestseller, offering an influential account of leading-edge scientific research. The pre-publication furore around this text was, in many ways, indicative since *The Double Helix* conjured a powerful image of modern scientific heroism, which was both highly controversial and very attractive.

Watson played with form and was influenced by various literary precedents. He apparently modelled the text on Sinclair Lewis' *Arrowsmith* ([1924] 1953) and the original title chosen for his book was *Honest Jim*, associating it with the popular Kingsley Amis novel *Lucky Jim* (1954). The published version of *The Double Helix* folded the trope of scientific discovery into the popular genres of the detective story and the memoir of youthful adventure. This meant that readers could engage with the text in a variety of ways. Watson provided technical detail about the research and represented his version of the scientific puzzle that could appeal to scientists (Stent 1980b). But a wider readership was offered a range of other ways into Watson's story. Following the generic route for readers of detective fiction, they could be carried through the twists and turns of the narrative, with Watson as sleuth, picking up the clues pertaining to the structure of DNA.

Ambition, controversy and concomitant broad popular appeal propelled this book and helped to make it an important and distinctive text in the history of science. Watson had indicated his concern to throw off the 'rather mystical' (Sayre 1975: 156–7) image of scientific discovery and to present a realistic picture of the doing of science. These sentiments cannot be taken at face value, nor can the trope of realism be invoked to explain or naturalize Watson's story of scientific adventure. Instead, my proposal is to delineate the features of James

Watson's heroism to gauge the significance of this portrait of the contemporary scientist and science. (I am referring here to the character represented in the text and not to the author/scientist himself.) Watson, in effect, projected an image of the quintessential modern, secular scientific hero and he thereby sketched a set of attributes which became associated with this late twentieth-century figure. The modern scientist was, according to Watson's vision, ordinary, sexy and racing, and the instantiation of these tropes constituted a new model of scientific heroism. It is my contention that *The Double Helix* is a crucial text in the history of modern natural sciences because of this powerful refiguring of scientific heroism. While some of these tropes have been highlighted by previous commentaries on this text, there has not been an exploration or evaluation of this dramatic reconfiguration and of the interplay of these elements. The first section (pp. 47–51) of this chapter outlines this refiguration, while the second section (pp. 51–67) considers the consequences of this casting of scientific heroism for the presentation of female heroes of twentieth-century science.

The scientist as ordinary guy

In a commentary about *The Double Helix*, Francis Crick referred to Watson's desire 'to show that scientists were human, a fact only too well known to scientists themselves but apparently not, at the time, to the general public' (Crick 1974: v). While Crick makes light of this facet of the book, it was a key feature of Watson's self-portrait: he presented himself as an 'ordinary guy'. Readers were thereby invited to identify with him and encouraged to see his youthful mishaps and misdemeanours as amusing and even endearing (even if some would find these annoying). It has been suggested that this image counteracted established popular images (from literature and film particularly) of the scientist as mad or bad (Haynes 1994; Van Dijck 1998; Frayling 2005).[2] But such representation also provided readers with points for imaginative identification: the world of science was made more accessible, more open, more likely to garner public interest and support.[3] Of course, as I shall detail on pp. 49–51, these imaginative positionings were also intensely gendered. Indeed, from an historical perspective Watson's tale proclaimed the secularization of contemporary science and its heroes, eschewing their clerical roots and associations (Noble 1992; McNeil 1997). Moreover, this alternative persona (alter ego) softened the image of the driven, competitive (or possibly frightening) scientific hero. The conjured portrait of a youthful Jim Watson, as an ordinary guy, resonated with readers attuned to the adventures of *Lucky Jim* (Kingsley Amis' hero) (Amis 1954), Holden Caulfield (hero of J. D. Salinger's *The Catcher in the Rye* [1951] 1958) and Jack Kerouac's *On the Road* (1957).

Secular, manly heroes

Jim Watson was a thoroughly modern hero and, like his literary counterparts, he flourished in a 'man's world'. Affiliating with Kingsley Amis' *Lucky Jim* and much of the popular adventure and detective fiction of the 1950s and 1960s, *The Double Helix* displayed and enacted the characteristic gender inequalities and heterosexism of the period. Science had long been established as 'a world without women' (Noble 1992), but Watson's story made modern science (and male scientists) sexy by exposing, celebrating and policing its modern heterosexist character. There were four key elements to Watson's enactment of the heteronormativity of science:

- contemporary heteronormative practices and conventions were portrayed as facilitating and supporting the careers and creativity of male scientists;
- homosocial dynamics (male friendship, collegiality and competitiveness) were displayed as providing the drive for scientific achievement and progress;
- scientific talent was aligned with masculine heterosexual interest and prowess;
- feminism and feminists (actual or fantasized) were represented as threatening the progress of science.

Heteronormative science

Two key female characters in Watson's story were his sister, Elizabeth Watson, and Odile Crick, the wife of his scientific partner, Francis Crick. Early in the narrative, Watson entertains the hope that Elizabeth's attractiveness will be an asset in his pursuit of the secrets of DNA. Having observed that Maurice Wilkins seemed to 'notice that my sister was very pretty', Watson speculated that 'if Maurice really liked my sister, it was inevitable that I would become closely associated with his X-ray work on DNA' (Watson 1969: 28–9). Watson's expectations were not realized, but this anecdote is apocryphal, amplifying somewhat ludicrously the expectation that heterosexual relations would bolster masculine scientific achievement.

Following a similar pattern, Watson represents Odile Crick as furnishing wifely support (despite Crick's extramarital flirtations) through the provision of wonderful meals, domestic comforts and pleasant socializing. Odile is shown as incapable of participating in science in any other way, since, as Watson explains, '[n]ot only did she not know any science, but any attempt to put some in her head would be losing a fight against the years of her convent upbringing' (Watson 1969: 63). Indeed, most of the women who appear in Watson's story are presented as servicing men of science in diverse ways, including through the provision of sexual entertainment and distraction by 'the popsies' (au pairs) and 'the girls' (undergraduates) around Cambridge. Elizabeth Watson is an indicative figure in her brother's tale as the only information provided about her pertains to her boyfriends, her fiancé, her husband and, in the coda, her children. Her main contribution to the plot (and to science) is her typing of the crucial article (which was to be published in *Nature*) in which Crick and Watson presented their research on DNA structure.

Science as a man's world

While heterosexual relations provide the backdrop and the maintenance framework for heroic scientific enterprise in Watson's narrative, he shows male friendship, collegiality and competitiveness as the dynamic factors in the shaping of modern science. *The Double Helix* is, in fact, a classic adventure story that revolves around the creativity of a male partnership. Watson also traces an almost exclusively male research network that sustains and challenges the dynamic duo. Watson does register some aspects of these gender restrictions in pointing out that King's College London, where Rosalind Franklin and Maurice Wilkins worked in the early 1950s, barred women from key locations for collegial exchange, including the senior common room. In this world of men, the spur of rivalry and competition is ever-present. In Watson's story, Linus Pauling, working at the University of California, is portrayed as the main rival in the 'race' to uncover the structure of DNA.[4]

Scientific heroes and heterosexual prowess

Although Watson highlights the homosocial aspects of leading-edge science in the second half of the twentieth century, he assures his readers that modern scientists have thrown off other elements of their vestigial clerical legacy (Noble 1992). Indeed, he provides ample (if not excessive) evidence that he and his colleagues are actively heterosexual. The book includes recurring references to Crick's attraction to young women and Watson's rather envious observations of his scientific partner's skills in flirtation. Watson also recalls ostensibly amusing or embarrassing encounters with other colleagues while they were engaged in heterosexual activities. Overall, *The Double Helix* conveys the image of the modern scientific hero as eschewing clerical celibacy, displaying instead the robust heterosexuality that became a feature of Britain's 'swinging 1960s'.

The feminist anti-hero

Watson's enthusiastic picture of the modern scientific hero as an ordinary guy who exuded conventional heterosexual masculinity was accompanied by a very different representation of Rosalind Franklin, who was the only woman shown actually doing science in the book.[5] Franklin is portrayed as difficult, closed to and in some ways obstructing the road to scientific progress, and, even, at one point, physically threatening. Watson also recounts that she was unfeminine and accordingly unattractive. Although this vituperative portrait of Franklin was subsequently contested and condemned (Sayre 1975; Yoxen 1985; Van Dijck 1998; Maddox 2002), Watson established a powerful trope that had lingering purchase: feminism as threatening to modern science.

Watson introduced Rosalind Franklin in a vividly evaluative passage:

By choice she did not emphasize her feminine qualities. Through her features were strong, she was not unattractive and might have been quite stunning had she taken even a mild interest in clothes. This she did not. There was never lipstick to contrast with her straight black hair, while at the age of thirty-one her dresses showed all the imagination of English bluestocking adolescents.

(Watson 1969: 20)

This passage shows Rosalind Franklin as subject to visual scrutiny, her appearance and, concomitantly, her femininity assessed by a dishevelled, ambitious junior researcher. Watson's insertion of 'by choice' suggests that Franklin was engaged in a concerted strategy of rejection of the norms of femininity. He brings together evidence of this deliberate flaunting of femininity, with a reference to 'blue-stocking adolescents'. Thus, Watson's introduction of Franklin explicitly links her with feminism. He then offers a crude and, by his own admission, rather spurious explanation of Franklin's feminine deficiencies by attributing her feminism to a pathological mother–daughter relationship:

So it was quite easy to imagine her the product of an unsatisfied mother who unduly stressed the desirability of professional careers that could save bright girls from marriages to dull men. But this was not the case. Her dedicated, austere life could not be thus explained – she was the daughter of a solidly comfortable, erudite banking family.

(Watson 1969: 20)

The phrase 'it was quite easy to imagine' provides the clue that what is displayed here is Watson's own imaginative landscape. Indeed, Sayre (1975) would later charge that Watson's book circulated a fictional version of Franklin. Sayre's claim is, in her terms, justifiable, but instead I prefer to draw attention to the broader imaginative landscape that sustains and shapes *The Double Helix*, in a way that unsettles the distinction between fact and fiction.

Later in the text, Watson reflects on Franklin's position in Maurice Wilkins' laboratory at King's College London, offering the assessment that 'she had to go or be put in her place' (Watson 1969: 20). Moving beyond this specific response to Franklin, Watson surmises that '[t]he thought could not be avoided that the best home for a feminist was in another person's lab' (Watson 1969: 21). His assessment of Franklin entails a summary of expectations regarding feminists – or, indeed, aspiring women more generally – in many scientific settings during this period. Watson's portrait of modern science suggested that women had to be kept in their place and that feminism threatened scientific progress. Towards the end of the book, Watson expresses relief when Franklin apparently accepts his and Crick's version of the double-helix structure. This prompts a striking reappraisal: 'Her past uncompromising statements on this matter thus reflected first-rate science, not the outpourings of a misguided feminist' (Watson 1969:

134–6). According to this evaluation, in supporting Watson and Crick's theory Franklin proved that she was a good scientist, rather than a feminist.

Racing, racy, competitive scientists

The third trope of Watson's imaginative construal of scientific heroism was speed – his heroic scientific adventure was portrayed as a 'race'. From the perspective of the early twenty-first century such a perception of leading-edge bioscience may seem so commonplace that it is difficult to register the significance of Watson's imaginative framework. However, it is striking that, in many respects, Watson portrayed a rather relaxed world that Edward Yoxen has characterized as a 'Golden Age' of scientific work 'when people had time to play tennis, go to the cinema and try somehow to make contact with members of the opposite sex' (Yoxen 1985: 178). José Van Dijck (1998) has unpacked some of the background to this trope, contending that the Cold War and the space race were influential in Watson's conceptualization of science. Moreover, an emphasis on speed and competition came to seem appropriate as pressures around funding for science became increasingly important in the United States and some other Western countries in the later decades of the twentieth century. As Yoxen notes, '*The Double Helix* was a tract on competition in science and the constant need to confront it' (Yoxen 1985: 178).

The identification of molecular biology with speed and competition became crucial to its emergence as the leading-edge of technoscientific development in the second half of the twentieth century. Yoxen explains:

> Having developed from the marginal pursuit of a few far-flung pioneers braving the disapproval of their peers, molecular biology became the site of some of the most intense competition in science and acquired the reputation as a field in biology where the real excitement lay and where the most daunting problems would continue to be found
>
> (Yoxen 1983: 43)

In fact, speed and competition have become dominant features of late twentieth- and early twenty-first century biosciences. In genetics and genomics (particularly during the lead-up to the 'completion' of the Human Genome Project in 2000) scientific heroes have been consistently represented as competitors in an international race. In this respect, Watson's image of racy, competitive scientific heroism has been thoroughly normalized.

Other lives in science

Watson told his own 'personal' story of scientific discovery and *The Double Helix* was, in many senses, a cocky tale, written in the aftermath of the award of a Nobel Prize. The other books considered in this chapter are biographies rather than autobiographies. Sayre, Goodfield and Keller undertook to tell stories of *other* lives in science.⁶ While many scholars have contended that these genres merge, signalling this by the use of the meta-category 'auto/biography' and through analytical work oriented around this fusion (Stanley 1992), the distinction in form is not insignificant in relation to these texts. *The Double Helix*'s specifically *auto*biographical form suggests both Watson's personal confidence and the broader cultural consolidation of men's place in Western science in the middle decades of the twentieth century which gave him 'voice', enabling him to tell his own story. In relation to Watson's text, the biographical form of the telling of the lives of these female scientists indicates both their more precarious positions within their scientific communities (discussed on pp. 53–7) and the more general profile of women within the natural sciences at this time, when it was difficult for women to have a life and a voice in this world.⁷

Nevertheless, as with Watson's tale, there were personal dimensions to all of these books. There was an acknowledged personal bond which propelled these projects and the authors became, in effect, public advocates for the scientists whose lives they researched. In Sayre's case, a strong personal friendship and sense of loss (in the wake of her friend's early death) sustained the author, expressed in the reflection '[t]o this day, I miss her' (Sayre 1975: 187). Sayre's admiration for Franklin inspired her to attempt to set the historical record straight on the scientist's achievement. Her identification was so intense that it led to her rhetorical claim that had the scientist lived longer 'she could scarcely have been overlooked' for the Nobel Prize (Sayre 1975: 188). For Goodfield, friendship and admiration emerged from, rather than being the sustaining background to, her study. It was the spark of an initial dinner-party encounter which launched the project: 'I knew that I wanted to follow Anna Brito's thought and work' (Goodfield 1982: 51). Keller had caught glimpses of McClintock around Cold Spring Harbor when the younger woman visited the laboratories there as a student in the 1960s and she was curious about McClintock's controversial scientific reputation. Keller writes of Barbara McClintock 'providing inspiration' (Keller 1983: xxv) and of a 'surprising warmth' and 'engaging ... personal exchange' (Keller 1983: 17) in her first interview with the scientist.

Like *The Double Helix*, all three of these books trace scientific discovery and document outstanding scientific achievement. Moreover, these authors convey a sense of mission regarding the standing of their subjects in their scientific communities. They claim to see what their subjects' scientific communities either had failed to see (Sayre), saw belatedly (Keller) or were yet to recognize (Goodfield). Their documentation thus helped secure their subjects' status within their scientific fields, as well as in a wider public context. Admiration for their subjects and high levels of identification are not uncommon amongst authors of biographies. Although, as Michael Shortland and Richard Yeo point out in a discussion of recent scientific biographers, such authors 'rarely explain[s] why a particular biography was written and why in a chosen style' (Shortland and Yeo 1996a: 31). The explicitness regarding intention and about personal bonds with their subjects are thus rather distinctive markers of these books.

While the early 1970s was the beginning of the age of the mass circulation paperback popular science text, biographies were firmly established as the best-selling non-fiction genre of the late twentieth century in North America and the United Kingdom. Second-wave feminism was a crucial factor sustaining ventures in women's scientific biographies. Feminists of this period were ambitious about the generation and circulation of stories of women's lives and eager to uncover the contribution of women in various spheres of culture. The feminist presses and women's bookshops which emerged in the late 1970s and 1980s in the United Kingdom, the United States and elsewhere catered to and fuelled the interest in women's stories. In this context, the documentation of scientific achievement undertaken by Sayre, Goodfield and Keller took on further significance, particularly given the chronic underestimation of women's contributions to the natural sciences. In fact, the ambition and significance of these cultural interventions were themselves symptomatic of that underestimation. Most biographical writing merely provides a gloss or fresh perspectives on the achievement and acknowledgement already accorded outstanding scientists. These books did more than this. The Franklin and McClintock biographies are linked to what Sandra Harding calls 'the "women worthies" project concerned with restoring and adding to the canons the voices of significant women in history' (Harding 1986: 30).

Women scientists: extraordinary lives

While Watson conjures the modern male scientist as an 'ordinary guy', particularly in relation to his sexual orientation and activities, these authors tell more complex stories of sacrifice, tension and denial, and of women who were continually cast as *extra*ordinary. In each of these studies the author suggests that her subject's entry into higher-level scientific education precipitated family conflict or withdrawal of support. Indeed, in setting out on a career in science each of the heroines is shown as defying gender conventions and expectations. Sayre indicates the ways in which Franklin broke with her family's expectations regarding young women's education and career trajectories. Indeed, Sayre provides a fairly detailed exposition of the elements of conventional femininity for women of Franklin's class in that period in Britain.[8] She outlines the ways in which Franklin deviated from these conventions in pursuing her career. Goodfield explains, in relation to Anna, that in Portugal 'at that time an only female child was expected to be a loving daughter, get married early, and produce grandchildren. She was not expected to become a medical doctor, reject her country, leave home and stay away' (Goodfield 1982: 7). Of course, such patterns of breaking with established trajectories were not atypical for young women from Britain, the United States and Portugal who entered science education in the early to middle decades of the twentieth century.

Keller presents McClintock as a figure who was clearly not ordinary in any of the ways Watson conjured. This is signalled in the introduction when Keller refers to McClintock's 'pursuit of a life in which "the matter of gender drops away'" (Keller 1985: xxv). It is also apparent when reference is made to McClintock's 'wish to be free of the body' (Keller 1985: 36). Nevertheless, the explanations here and the subsequent unfolding of Keller's scientific career establish that she did invest extensively in physical observation and other corporeal activities. Keller demonstrates that this disavowal clearly pertains to McClintock's desire to throw off contemporary expectations about feminine appearance and conventions around gendered embodiment.

Women in science: heteronormative restrictions

James Watson exposed, enacted and celebrated the heterosexist gender dynamics that sustained his life in science and that of his male colleagues. It was only in the Epilogue to *The Double Helix*, which was a controversial apology for his negative portrayal of Rosalind Franklin that Watson reflected on the significance of these dynamics for female scientists. Here he refers to he and Crick 'realizing years too late the struggles that the intelligent woman faces to be accepted by a scientific world which often regards women as mere diversions from serious thinking' (Watson 1969: 133). Nevertheless, Watson had himself put into circulation a powerful popular narrative about modern science which effectively re-enacted precisely the kind of denigration of women he discussed in his Epilogue.

The three biographies considered in this chapter cast the gender relations of modern science in a very different light. They show that for these female scientists Western heterosexist normativity, far from facilitating their lives in science, produced tensions and conflict. These lives in science are shown to be built around confrontation and difficult negotiations with, rather than flamboyant enactments of, heterosexual norms and expectations. While Watson recounts how heterosexual conventions *support* male careers and creativity in science, the biographers of these female scientists show a very different set of dynamics. So, for example, Sayre contended that Franklin 'sacrificed' having children to facilitate her scientific career, explaining that '[t]his was what she gave up as the token and sign of her sincerity and her commitment' (Sayre 1975: 53–4).

While Goodfield quotes Anna rejecting Sayre's framing of female scientific life involving 'sacrifice', she indicates various turning points in Brito's 'life as a woman' (Goodfield 1982: 65), showing that this was often in tension with her life as a scientist. Indeed, Goodfield's (and Anna's) conceptualization of her 'life as a woman', as distinct from her 'life in science', contrasts sharply with Watson's representation of the smooth complementarity between sexual and scientific identities amongst his male scientific colleagues. In some instances Goodfield is explicit about resulting conflicts. For example, she recounts an early period of the scientist's career: 'It was a time of great tension. Anna was contemplating marriage and was generally expected to assume a woman's traditional role. The conflict between these claims and those of a concentrated scientific career strained her to the breaking point' (Goodfield 1982: 34). At other junctures she merely hints at tensions and difficulties, as when she quotes Anna as observing, in an important moment in her career, that 'many roads have come to an end with my life as a woman too' (Goodfield 1982: 145). At another juncture Anna reflects:

> The woman in me wants to give up, but the total person I am is here, and rejoices in Michael's [a research colleague's] voice telling me that ferritin-positive cells and transferring-positive cells are in the thymus-dependent area of the lymphatic spleen.
>
> (Goodfield 1982: 152)

This splitting of Anna's identities as woman and as scientist is in stark contrast with Watson's vision of the enactment of masculine heroism in and through science. Keller explains that McClintock copes by withdrawing from the obvious conventions of femininity in the hope that ' the matter of gender' 'drops away' (Keller 1983: 26). In diverse ways, these biographies show, in contrast with Watson's story, that heterosexual norms did not work positively for twentieth-century women working as scientists.

Homosocial networks and female exclusion

The homosocial networks which sustained Watson and his colleagues and which were reinforced by institutional structures figure rather differently in these biographies of female scientists. Sayre shows in considerable detail the consequences of prohibitions about women's full participation in academic life at King's College London in the 1950s. She also indicates how more informal networks marginalized Rosalind Franklin as a female scientist (Sayre 1975: 98). She characterizes Franklin's experiences of these collegial limitations as a kind of 'purdah' (Sayre 1975: 97). Keller sketches the restrictions which, in the 1930s to 1950s in the USA, channelled McClintock and other women scientists away from research and high-level posts. In a chapter strikingly titled 'A Career for Women' (Keller 1983: ch. 4), she explains how McClintock encountered the traps and limitations of the gender segregation that operated in the natural sciences in mid-twentieth-century USA: 'women in the sciences tended to be scientific workers and teachers rather than scientists. ... Careers as research scientists were not available to them' (Keller 1983: 52). Although each of these authors shows their female subjects working effectively with other colleagues and students as well as becoming valued mentors, all three women emerge as relatively isolated and as marginalized in the male-dominated collegial networks of their scientific fields.[9]

Scientists and sexuality: 'men act and women appear' (Berger 1972: 47)

> To be the object of vision, rather than the 'modest,' self-invisible source of vision, is to be evacuated of agency.
>
> (Haraway 1997: 32)

While James Watson flaunts the sexual prowess of the men of science around him, it is striking that there is no explicit reference to sexual relationships in these accounts of the lives of these outstanding female scientists. Sayre notes that, during her visit to the USA in 1956, Rosalind Franklin 'met a man whom she might have loved, might have married' and that 'she put this out of her mind, but she went on living, fierce and even passionately' (Sayre 1975: 184). Goodfield (1982: 34) indicates that Brito considered marriage before embarking wholeheartedly on her career. In these stories of high-achieving twentieth-century women scientists, sexual frisson is not linked to scientific achievement and, in this respect, these women are cast as lonely figures. Whatever the nature of their actual lives, their sexual activities were not represented as relevant to or enhancing their scientific careers or status.

Watson's celebration of the sexual prowess of his male scientific colleagues contrasts sharply with Sayre's and Keller's accounts of the suspicion garnered about the femininity of Franklin and McClintock. They demonstrate that these scientists were continually subjected to the critical male gaze. Keller traces in the files of the Rockefeller Foundation (the most important US scientific funding agency in the twentieth century) a note written by a key administrator which records that 'Miss McClintock has a slight, boyish figure, weighing about 90 lbs., with a tousled boy's haircut' and 'that her father was so greatly disappointed that she was not a boy that he proceeded to raise her as a boy'. This note summarily concluded: 'And she still looks and acts more boy than girl' (Keller 1983: 75). As noted previously, Sayre registered the creation of what she regarded as the fictional character 'Rosy'. Watson's character was, in Sayre's estimation, 'the perfect, unadulterated stereotype of the unattractive, dowdy, rigid, aggressive, overbearing, steely, "unfeminine" bluestocking, the female grotesque' (Sayre 1975: 19). She continued: 'By choice, he tells us, she refuses to emphasize her feminine qualities – by which he means (and it is his idea of femininity) that she is badly dressed, wears no lipstick, does nothing interesting with her straight black hair' (Sayre 1975: 21).

Sayre and Keller tackle these negative appraisals in rather different ways. Nevertheless, they both offer more positive assessments of these scientists' appearance. Sayre contests Watson's version of Franklin as a distortion and she rebuts specific elements of his portrait: 'People with whom Rosalind worked in both England and in France thought her rather smart, always well-groomed, discernibly English in her style, but far from habitually dowdy'. Moreover, she contends that 'the lipstick was almost invariably there' and that Franklin did not ever wear spectacles, explaining: 'Rosalind had the eyesight of an eagle, and resorted to magnifying lenses only for the closest of fine work' (Sayre 1975: 21). Indeed, Sayre's efforts to document evidence of Franklin's femininity were extensive and disturbed at least one of the reviewers of this book.[10] Keller, by contrast, does not directly contest the appraisals of McClintock's appearance as unfeminine. Instead, she makes this a matter of individual style, giving it a positive gloss by attributing it to a distinctive personal, aesthetic orientation. She notes:

Her slacks and shirt pointedly rejected feminine fashion, but they were carefully pressed. The economy of her words and movements, the way she dressed, the way she moved and talked – all expressed a fastidious spareness, an aesthetic of order and functionality.

(Keller 1985: 17)

These contestations draw attention to the ways in which the ubiquitous evaluations associated with the male gaze were enacted with distinct consequences for twentieth-century female scientists. Moreover, these authors demonstrate that they were aware that the focus on appearance detracted attention from the scientific activities of these women. They take issue with negative appraisals in quite different ways: Sayre assembling evidence of Franklin's feminine appearance to discredit Watson's description; Keller following McClintock's lead in eschewing any reference to gender by emphasizing her androgynous 'aesthetic of order and functionality' (Keller 1983: 17). Nevertheless, these reassessments of appearance and sartorial tastes themselves enact the very modes of appraisal that were a crucial and problematic part of the intensely binarized, heterosexist regimes of the scientific worlds their subjects inhabited.

The threat of feminism

Watson's reported reappraisal of Sayre when she accepted the double-helix hypothesis was stark: she had proven that she was a good scientist rather than a feminist. In *The Double Helix* feminism was presented as threatening Watson and his colleagues, and the very activity of science itself, although Watson never explains *why* he finds it so menacing. This construal of the relationship between science and feminism casts its shadow over the three biographies considered in this chapter. On the one hand, all three biographers register the gender inequalities they have observed in the world of science and its consequences for their subjects. Nevertheless, they also demonstrate that these scientists were *not* feminists.

Sayre traced what she considered to be the mistaken identification of Franklin with feminism. According to Sayre, this begins with Franklin's PhD supervisor, R. G. W. Norrish, and recurs most persistently in Watson's portrayal of her as a 'bluestocking'. In dismissing Norrish's perception of Franklin, Sayre insists that the female scientist was '"[f]eminist"only in the widest philosophical sense, not in an activist one' and that her position 'had little in common with doctrinaire, or political, feminism ... for the simple reason that it was not fundamentally feminist' (Sayre 1975: 58–9). Instead, Sayre noted that Franklin displayed an 'attitude ... of exacting professionalism' (Sayre 1975: 59). Somewhat contradictorily, in the 'Afterword' of the book, Sayre does express concern that Franklin 'had been taken up, in the Watson version, by antifeminists everywhere' (Sayre 1975: 197).

There is only one direct reference to feminism in Goodfield's book, despite the accounts of the tensions precipitated in Anna's 'life as a woman' by her life as a leading-edge scientist. At one point, Anna reflects: 'Mine is the eternal problem of women in science. Very, very few women bring this off. But it is our own choice; it is self-inflicted.' She then reassesses the gender specificity of her situation:

> This is, of course, the eternal problem of anyone in science. It yields its secrets only to single-minded obsession, and at the end of the lonely day, the week, the year, or the decade one emerges from the laboratory into what, humanly speaking, is a very empty space.
> (Goodfield 1982: 149)

In her final musings about scientific creativity, Anna is reported as recalling a revealing episode in which she was perceived to be a feminist. She reflects that

> the conception of a concept, or the gestation of an idea, is really a kind of birth. It is the only time when men can share with women an essentially female experience; the only time when a man can experience anything like giving birth.
> (Goodfield 1982: 230)

She states that she 'said this to a man in England, he was frightened ... or annoyed. He thought I was saying it for feminist reasons. But I wasn't' (Goodfield 1982: 230). Here Brito rehearses the argument which emerged in feminist and critical science studies scholarship of the 1980s (Easlea 1980a, 1983; Merchant 1980)[11] that scientific creativity was, in some ways, a substitute for feminine reproductive creativity. However, she quickly denies any feminist intentions or identity, disavowing the feminist connotations of her framing of science.

In the foreword to Keller's book, Rollin Hotchkiss declared that as 'a humanist, rather than a feminist' McClintock 'expected unprejudiced respect for herself and other women' (Keller 1983: xiii). In the acknowledgements, Keller puts great store in McClintock's 'rejection of female stereotypes' and the 'pursuit of a life in which "the matter of gender drops away"' (Keller 1983: xxv). She explains that

> she was adamant; she was too different, too anomalous, too much of a 'maverick' to be of any conceivable use to other women. She had never married, she had not, as an adult or as a child, ever pursued any of the goals that were conventional for women. She had never had any interest in what she called 'decorating the torso'.
> (Keller 1983: 17)

She notes that McClintock 'had come to insist on her right to be evaluated by the very same standards as her male colleagues' (Keller 1983: 76–7).

In a later important reflection on gender relations in science in the USA in the twentieth century, Evelyn Fox Keller observed:

Throughout this [the twentieth] century, the principal strategy employed by women seeking entrance to the world of science has been premised on the repudiation of gender as a significant variable for scientific productivity. The reasons for this strategy are clear enough: experience had demonstrated all too fully that any acknowledgement of gender-based difference was almost invariably employed as a justification for exclusion. Either it was used to exclude them from science, or to brand them as 'not-women' – in practice, usually both at the same time. For women scientists *as scientists*, the principal point is that measures of scientific performance admitted of only a single scale, according to which, to be different was to be lesser. Under such circumstances, the hope of equity, indeed, the very concept of equity, appeared – as it still appears – to depend on the disavowal of difference.

(Keller 1999: 236)

Keller's observation resonates with the accounts of the lives in science registered in these books. There is ample evidence provided to indicate that each of these scientists repudiated gender 'as a significant variable for scientific productivity'. Sayre asserts that Franklin operated with the 'assumption that rational people would easily understand without further demonstration that she deserved to be judged not as a woman scientist, but as a scientist pure and simple' (Sayre 1975: 54). Keller contends that McClintock wished 'to transcend gender altogether' (Keller 1983: 76) and that, as her career developed, '[i]n effect, she was refusing to accept a woman's place. ... She had come to insist on her right to be evaluated by the very same standards as her male colleagues' (Keller 1983: 76–7). Anna repeatedly rejects contentions that gender mattered in science. She regards Sayre's account of Franklin 'sacrificing' children for her career as a distortion (Goodfield 1982: 65) and she quickly reframes her own construction of 'the eternal problem of women in science' as 'the eternal problem of *anyone* in science' (Goodfield 1982: 149).

Nevertheless, as Keller and the other authors of these biographies clearly establish, 'the matter of gender' did *not* 'drop away' and these books demonstrate why it mattered in these women's lives in science. These texts have been written in the wake of second-wave feminism and, in different ways, the authors were influenced by this movement. They offer observations of gender divisions, inequalities and forms of discrimination with which each of these female scientists contended. Keller observes that McClintock learned that '[n]o efforts of her own would erase the fact that she was a woman in a profession institutionally established for men' (Keller 1983: 76). Summing up Franklin's career, Sayre concludes: 'She was a very good scientist and a very productive one, a very honest one of unimpeachable integrity'. Alluding to the problems gender relations had created for Franklin (which she delineates in the book), she then interjects: 'she was not the less of any of these things because she was a woman, and often opposed on no better grounds than her sex' (Sayre 1975: 197). Although Goodfield does not explicitly highlight specific gender barriers inhibiting Anna once she embarks on her career, as noted previously, she does expose the tensions Anna experienced between her life as a woman and her life as a scientist.

The long quotation from Keller given above suggests that in the twentieth century female scientists were likely to be regarded as suspect both as scientists and as women. While these three popular texts on women's lives in science are primarily oriented towards convincing a wide public readership that their subjects had made important contributions to the twentieth-century biosciences, they manifest an awareness of the double scrutiny to which these women had been subjected. Nevertheless, they dispute (Sayre 1975) or contain (Keller 1983) previous deprecating assessments of these scientists' unfeminine appearance. In addition, both Sayre (1975) and Goodfield (1982) provide evidence of their subjects' heterosexual credentials through reference to marriage possibilities. Hence, whilst contesting some specific evaluations, in effect they reinstantiate the dual assessment mechanisms which circumscribed the lives of women scientists during this period.

For Watson, the feminist was the figure who posed the greatest threat to science. Not surprisingly, these accounts of women's scientific achievement, which followed in wake of *The Double Helix*, trod cautiously around issues of both femininity and feminism. Feminism is a spectral presence in all of these texts: as I have indicated, it figures explicitly almost exclusively in disavowals and distancing. Yet the terms of reference and interpretation deriving from second-wave feminism inform these representations of women's lives in science. Moreover, it was the late twentieth-century Western feminist interest in 'women worthies' and in rewriting the canon of Western cultural achievement which helped to create an enthusiastic readership for these texts. Nevertheless, it was only with the appearance of Donna Haraway's *Primate Visions* (1989: esp. 277–367) that feminism came fully out of its scientific closet. Haraway's analysis of the making of primatology as a new science of the twentieth century presented profiles of four 'North American white women' (Haraway 1989: 303) – scientific heroes – openly working at 'the intersection of feminism and the science of monkeys and apes since about 1970' (Haraway 1989: 285–6).

Out of the race: doing careful science

The four books considered in this chapter provide narratives of discovery. In each case, the scientific discovery (of the double helix in Watson's and Sayre's texts, of the behaviour of lymphs in cancer in Goodfield's, of gene transposition in Keller's) is the fulcrum for and climax of the life-story presented therein. In this sense, they are conventional stories of scientific heroism.[12] In Watson's case, the achievements associated with the discovery of the double helix had already been celebrated through the award of the Nobel Prize in 1962. Thus, the book provided a retrospective account of the activities through which he (and Crick) had 'earned' this prestigious international commendation. The exclusive focus on this episode, without reference to the preceding history of genetics or to subsequent developments in the period between 1953 and the publication of

Watson's book in 1968, contributes further to the heroic casting of Watson and Crick (Van Dijck 1998: 38–45). The narratives of discovery which Sayre, Keller and Goodfield constructed did not have such firm cultural underpinning. These authors had to establish their claims regarding heroic accomplishment without the benefit of preceding public cultural legitimation.[13] In addition, Sayre and Keller tried to address and dispel negative images that had accrued to their subjects within their scientists' own research communities. More generally, these three biographies demonstrated to a wide reading public that these figures merit international renown.

These claims for scientific heroism, through tales of discovery, also incorporated explorations and contestations concerning how science could and should be done. As I have indicated previously, Watson's was an irreverent account which highlighted masculine camaraderie, competition, flashy insight and happenstance, although model building was the scientific method which was lauded in his story. The three biographies presented in this chapter offer very different visions of the *doing* of science, explicitly or implicitly contesting Watson's vision of racy science.

Sayre took direct issue with many aspects of Watson's picture of scientific research, which she feared would be excessively influential because of the popularity of *The Double Helix*. She contrasted Franklin's systemic crystallographic approach to the investigation of DNA with Watson's impulsive model building, asserting the importance of traditional empirical methods and objective reasoning (Sayre 1975: 146) in science. However, methodological issues were not her only concern in relation to the doing of science. She raised questions about ethical codes of conduct pertaining to the ownership and circulation of research materials and findings. Moreover, she expressed her scepticism about the ethos of racing, suggesting that, in accordance with her image of Franklin, good scientific research required slow, careful observation and experimentation.

The prologue of *An Imagined World* consists of one of Anna's letters laying out the domestic detail ('a corridor of smells ... cages of mice ... surgical instruments ... petri dishes ... sterile bottles of medium') of her laboratory, with her declaration: 'this is what science is made of'. This image and declaration convey a starkly different orientation to scientific research to that provided in Watson's narrative. The text enacts an ongoing dialogue between Brito and Goodfield about scientific research in which Brito repeatedly rejects Goodfield's textbook history and conventional philosophy of science vision of how science works. The scientist insists on the network of agents and elements involved in the making of science. The human dimension of such networks which is foregrounded is teamwork, which for Brito also included the contributions of technicians and cleaners. Information about the economic and bureaucratic structuring of research science peppers the account, particularly when Anna is shown struggling with the constraints of the US research grant system.

While Brito's extended and rather domesticated model of scientific creativity is fully registered, the text is also shaped by Goodfield's interest in the psychology of outstanding individual scientists. The discussions and arguments recounted in the book carve out a model of personal motivation in scientific investigation that is far removed from Watson's picture of male bonding through competition. Anna contends that 'the best analogy' for scientific research 'is always love – making love' (Goodfield 1982: 69). Parallels are established between artistic and scientific creativity, through Anna's wide-ranging high-cultural references to visual artists, musicians and poets. Beyond this, she insists that empathetic identification with elements of the natural world sustains good scientific research, commenting: 'If you really want to understand about a tumor, you've got to *be* a tumor' (Goodfield 1982: 226).

Keller (1983: 207) quotes this guideline, thereby establishing the affinity between the Portuguese researcher and McClintock in their orientation to scientific research. Keller documents McClintock's years of dedicated observational research and emphasizes that the cyto-geneticist cultivated and was sustained by her 'feeling for the organism'. As Keller (1989) emphasized in a later commentary about this biography, she makes the case for the enrichment of science through diversity in methodological and personal approaches through McClintock's story. Hence, McClintock's 'naturalist's approach' (Keller 1983: 207) is posed as unsettling the hegemony of the structural, molecular biology which had been instantiated, in part, through Watson's tale of the revelation of the double-helix structure of DNA. Similarly, Keller's observation that, for McClintock, 'years of close association with the organism she studies, is a prerequisite for her extraordinary perspicacity' (Keller 1983: 198) indicates that the pace of scientific achievement may be far slower than Watson's memoir suggests. Here and in her elaboration about McClintock's mystical engagement with the natural world she emphasizes the bonding between scientist and nature, rather than the collegial male bonding which Watson celebrates as a feature of leading-edge scientific research.

Receptions and refractions: up close and personal

The conjurings of scientific heroes reviewed in this chapter were highly personalized projects through which each of these authors established the position of their subjects in their scientific communities while seeking to enlighten a broad lay readership about how science works. These authors claim insider status: Watson as one of the main actors in the discovery of the structure of DNA; Sayre as a close friend and wife of one of Franklin's colleagues; Keller as a former scientist and historian of science; and Goodfield as an established historian of science and a science journalist.[14] As previously indicated, strong personal bonds between the authors and their subjects characterized and sustained the biographies of women scientists analysed in this chapter.

However, by the time Evelyn Fox Keller's biography of McClintock appeared in 1983 a new professional social studies of science was making its mark by offering new kinds of 'laboratory lives'. The latter term is an adaptation of the title of Bruno Latour and Steve Woolgar's influential study (*Laboratory Life: the social construction of scientific facts*, 1979),[15] which was one of the first and most lauded in this cluster of ethnographically based studies that came to characterize a powerful strand of social studies of science.[16] Anthropological in orientation, the authors of these books entered their field (laboratories) and studied the 'native tribes' (as they sometimes labelled them) and practices therein by rendering them strange and foreign. Suspicious of biographical accounts, heroic discovery narratives and other humanist tropes, these scholars generally produced abstracted, detached analyses of scientific work that decentred human agency. Their relationship to the kinds of projects analysed in this chapter is exemplified in Latour and Woolgar's dismissive characterization of Goodfield's *An Imagined World* as a 'detailed study of an individual scientist's experiences, … which fails to address the social process of laboratory work' (Latour and Woolgar 1986: 285, n.4.).

However, while they were determinately uninterested in the personal, these new social studies of science researchers did seek to position themselves close to canonical sites in the making of science – laboratories. Strikingly, their selection of other resources to bolster their studies resulted in some strange judgements. For example, Latour and Woolgar treat *The Double Helix* as a premier resource for their sociological analyses, one that they consider is not mired by the psychological preoccupations and humanist myth-making they seek to avoid. Citing a particular incident in Watson's narrative, they explain that

> Watson's portrayal of his 'pretty model,' in which bases are paired along a like-with-like structure, does not situate himself in a realm of thought, but inside a real Cambridge office manipulating physically real cardboard models of the bases. He does not report having ideas, but instead emphasises that he shared an office with Jerry Donohue. … If Watson had not written his book, no doubt the complexity of this practice would have been transformed, either into an anecdote that 'one day Watson got the idea of trying the keto form' or into a titanic epistemological battle between rival theories.
>
> (Latour and Woolgar 1986: 171–2)

This appraisal of *The Double Helix* as providing an untainted, non-idealized, materialist account of science in the making is elaborated in Latour's *Science in Action: how to follow scientists and engineers through society* (1987). In this book, Watson's highly personal memoir is treated as a primary source and reference text in Latour's mapping of strategies appropriate to the new social studies of science. In treating Watson's text as an unmediated account of science (as showing how it *really* works), Latour and Woolgar failed to acknowledge the ways in which *The Double Helix* performatively enacted its own mythical version of scientific achievement.

The biographies of women scientists examined in this chapter found a more receptive readership outside the developing professional science studies community. One reviewer noted that Keller's biography of McClintock '[u]nderstandably … became something of an inspiration for women working within science and technology, feminists and non-feminists alike' (Grobicki 1987: 211). As noted previously, auto/biography was a generic mainstay of second-wave feminism and the pursuit of 'women worthies' (Harding 1986: 30) was a crucial strategy in the challenging of established canons in all Western cultural fields. These accounts of lives in science were attractive to feminists involved in a social/political movement that took as its touchstone the slogan that 'the personal is political'. Moreover, a new feminist science studies was emerging as an interdisciplinary academic field during this period, including among its key figures Donna Haraway (1985), Sandra Harding (1986) and, indeed, Evelyn Fox Keller (1985) herself. Nevertheless, some feminist reviewers were dissatisfied with these biographies because they felt they did not offer sustained critical perspectives on science and gender norms (Hubbard 1976; Grobicki 1987). In addition, the biographies examined in this chapter intersected with two key foci of feminist controversy during this period which revolved around the possibility of distinctive women-centred or feminist forms of culture. These were debates about an identifiably feminine or feminist ethics of care[17] and about the prospects for a distinctive feminine or feminist science.

In 1991, Sandra Harding observed: 'the question "Can there be a feminist science?" has been raised by virtually everyone who participates in or contemplates the feminist discussions of the sciences' (Harding 1991: 296).[18] This preoccupation undoubtedly influenced some readings of the popular biographies of women scientists analysed in this chapter. Indeed, Evelyn Fox Keller explicitly entered into the debate about a 'feminist science' in the wake of responses to her biography of McClintock. Keller disabused interpretations of McClintock as the harbinger of a distinctive feminine or feminist science on a number of grounds, including that '[t]o ask women scientists to accept the notion of a different science representing a different reality ... would be to ask them to give up their identity as scientists' and that this would 'reinforce the traditional opposition between women and science' (Keller 1999: 240).

Nevertheless, Keller was subsequently charged by the science studies scholars Evelleen Richards and John Schuster (1989a, 1989b) with both advocating a feminist science and uncritically employing a methods discourse. Taking Keller's representation of McClintock's scientific practice and Sayre's account of Franklin's mode of scientific research as their main case studies, Richards and Schuster rehearsed the claims for 'social constructivist and contextualist analyses of scientific practice' (Richards and Schuster 1989b: 729) which they endorsed as preferable to 'the story of invention, refinement and deployment of the method' (Richards and Schuster 1989b: 727). The focus of the dispute between Keller and Richards and Schuster was the status of methods discourse, with Keller claiming its importance for scientific research and Richards and Schuster maintaining that it rhetorically glossed the actual activities of science, yielding only a mythical version of such research.

The laboratory studies of Latour and Woolgar and other ethnographers of the 1980s and the heated exchanges between Keller and Richards and Schuster were harbingers of the burgeoning new social studies of science. This new disciplinary field was forged through the development of distinctive anthropological modes of analysis and through critical renunciations of some traditional tropes of the history and philosophy of science. Rejection of humanist frameworks manifested in preoccupation with the scientific mind and individual scientists and of methods discourse were some of the key markers of this new 'post-Kuhnian' (Richards and Schuster 1989b) science studies.

The excitement around these innovations masked some conservative features of this new orientation in science studies. First, there was a reverential celebration of the laboratory which was treated as the main location for the making of science.[19] Second, in some of this work there was a tendency to identify and interrogate only *certain* forms of myth-making in accounts of scientific research (as Latour and Woolgar's 1986 uncritical use of *The Double Helix* illustrates). Third, the new orientation was manifested in a reluctance to discuss issues which pertained to more macro features of science as a social activity. The most obvious instance of this last pattern was the neglect of attention to the gender relations of science and the consequent failure to investigate the production and reproduction of science (at its highest levels) as 'a world without women' (Noble 1992; Haraway 1997). The abandonment of humanist approaches did yield a more complex, less anthropocentric picture of agency in the making of science. However, there was little incentive to develop more sophisticated understandings of human agency in scientific development.[20] Finally, the preoccupation with traditional sites and locations meant that the new social studies of science never fully acknowledged the diffused and multiple locations in which science was made.[21] In this sense, it was largely disconnected from the new sub-field of public understanding of science which emerged in the 1990s and which lacked the theoretical sophistication of its sister field.[22]

Meanwhile, as Christopher Frayling has recently argued, in the context of his review of cinematic representations of scientists, 'in the *public* rhetoric of science, biography is still as important as it has ever been' (Frayling 2005: 179). This chapter has analysed a crucial phase in the making of the modern scientific hero. It has traced the emergence of a new secular vision of the heroic scientist in the most popular scientific memoir of the twentieth century, *The Double Helix* (1968), which celebrated the heteronormative framing of science. Through the figure of Watson a distinctive set of tropes was established: the heroic scientist was an ordinary guy, competitive and racing, and profoundly threatened by feminism. The popular biographies of women scientists considered here take their cues from this new figuration of scientific heroism. They show extraordinary women, rather than the ordinary guy, and the other – troubled – side of the heteronormative conventions which Watson celebrated. They also describe diverse modes of and paces for doing outstanding science. Feminism was a crucial resource in the construction of these 'women worthies', yet it was kept at a distance. Meanwhile, a new social studies of science was forged with and against these popular books, while the new sub-field of public understanding of science emerged as a distinctive research as well as policy field.

Steven Shapin and Simon Schaffer's important science studies text *Leviathan and the Air-Pump: Hobbes, Boyle and the experimental life* (1985) conjured the figure of the scientific witness as the legitimated agent in the making of the modern experimental sciences. They trace the historical circumstances for the emergence of this figure in the social and political culture of seventeenth-century England. As they explain, science required 'a *modest man*' 'whose narratives could be credited as mirrors of reality ... his reports ought to make that modesty visible' (Shapin and Schaffer 1985: 65). Donna Haraway (1997) has, in turn, interrogated Shapin and Schaffer's own conjuring, recasting their study as an investigation, not only of the making of modern science, but also of the making of modern gender relations.[23]

Haraway maintains that there have been 'practical inheritances which have undergone many reconfigurations but which remain potent' from Boyle's version of the 'modest witness' and that 'the important practice of credible witnessing is still at stake' (Haraway 1997: 33) in contemporary science. The analysis of twentieth-century heroes of science offered in this chapter provides some perspective on twentieth-century reconfigurations of this model of scientific heroism. In fact, the furore surrounding the publication of *The Double Helix* (Stent 1980a) may have indicated fears that this idealized model of how science works would be undermined by the circulation of Watson's narrative. The contention of this chapter is that Watson's text does not banish the figure of 'the modest witness'; rather it offers a new alignment of heterosexual masculinity and scientific prowess – a new twentieth-century model of the heterosexist, virile 'modest witness'.

In contrast, the tales of lives lived in and through twentieth-century science reviewed in this chapter show the difficulties European and North American women encountered when they claimed to be 'objective, modest witnesses to the world' (Haraway 1997: 32). Some two centuries after Boyle's establishment of the social technology for experimental philosophy, in Europe and North America science was, indeed, more open to women (Rossiter 1982, 1995). The female scientific heroes of this chapter had gained high-level scientific education and, as qualified and accomplished working scientists, they were *positioned* as 'modest witnesses'. Nevertheless, they were repeatedly subjected to two forms of delegitimizing ploy. Brought under the scrutiny of the male gaze, they were made 'the object of vision' in ways that undermined their claims to be 'the "modest," self-invisible source of visions' (Haraway 1997: 32). In addition, the label 'feminist' was mobilized as a powerful disqualifying epithet – which proclaimed political perspective, rather than objective witnessing. These were the modes through which individual twentieth-century female scientists were delegitimized as scientific witnesses, despite their being highly professionally qualified and accomplished working scientists.

These popular biographies of scientists explored the 'critical boundary between watching and witnessing, between who is a scientist and who is not, and between popular culture and scientific fact' in the twentieth century (Haraway 1997: 33). While they tell very distinctive stories, the heroes of these tales are each cast with reference to two spectral figures – the feminist and 'the modest witness'. Watson vividly and insistently portrays these figures as antithetical. Against this background, and as explained on pp. 57–60, in the other texts considered here disabusing the label of feminist becomes a crucial element in sustaining claims regarding the scientific heroism of these twentieth-century women.

Although the hero of *The Double Helix* was a far cry from 'Boyle's celibate, sacred-secular, and non-marital man', he still claimed Boyle's legacy as 'a modest witness ... of the mind' (Haraway 1997: 32). The heteronormative and homosocial scaffolding of twentieth-century science were foregrounded and celebrated in Watson's tale of scientific achievement. His twentieth-century scientific hero was much more overtly and aggressively heterosexist and masculine than his seventeenth-century predecessor. Strikingly and somewhat ironically, the watchword that seems to dominate of the stories of Franklin, McClintock and Brito considered here is also modesty. However, these are tales of far more pervasive modesty – identified with these women's bodies, their aesthetics, their dealings with others, as well as with their personal styles in doing science. These biographies have been written with a great deal of passion and, on one level, this almost excessive emphasis on modesty may be indicative of these authors' efforts to gain cultural legitimacy for their subjects. Beyond this, it is also a reminder of the considerable work (including symbolic work) required in order to make women and persons of colour 'count as objective, modest witnesses to the world' (Haraway 1997: 32).

A postscript: 'post-feminist' recastings of scientific heroism

As the preceding analysis suggests, feminism was the ghostly presence which haunted, informed and sustained the biographical projects of Sayre, Goodfield and Keller. Rosalind Franklin and Barbara McClintock became heroines of second-wave feminism. Early in the twenty-first century, new biographies of both of these scientists appeared. Nathaniel Comfort (2001) took issue with many aspects of Keller's account; in particular, he questioned the portrayal of McClintock as a scientific outsider who had been kept at the margins of the scientific establishment. Comfort contends that Keller's reliance on interviews with McClintock and on a set repertoire of stories from the scientist produce a misleading picture which renders McClintock a victim rather than, as he presents her, as a main player in the history of genetics (Delamont 2005: 494–5). Brenda Maddox's (2002) volume, in contrast with Sayre's biography, offers a much more dispassionate account of Rosalind Franklin's life. The later biographer is much more matter of fact about Franklin's achievements, her personal strengths and weaknesses, and her sexual activities.

These are revisionist recastings that render McClintock and Franklin scientific heroes for the twenty-first century. Although a more detailed comparison would be required to detail the differences between the late twentieth- and early twenty-first-century versions of these scientific heroes, it is striking that these new biographies tell biographical stories in which the struggles and difficulties associated with gender are denied or contained. While feminism haunted the accounts of the lives of scientists analysed in this chapter, it is effectively banished in these revisionist versions.

Part II
Telling stories

5 New reproductive technologies
Stories of dreams and broken promises

> Infertility is considered such a good story these days.
> (Carpenter 2006: 41)

New Reproductive Technologies (NRTs) do not feel so new to me since they have been a focus of my research and reflection since the mid-1980s. Likewise, readers and viewers of Western media are less easily shocked by reports of the practices, the complex kinship patterns (see Franklin 1992; Haimes 1992; Strathern 1992c) and other consequences of relatively new modes of procreation in the early twenty-first century. Indeed, genomics and genetic engineering have rather usurped the place of NRTs as the strand of recent biotechnology which now musters most controversy and concern.[1] Nevertheless, NRTs have been kept in the public eye in recent years as they have generated countless new stories in the Western mass media which range from controversies about frozen embryos to 'sibling replacement' or to the reports about 'Britain's oldest mum' (see note 40).

This chapter began as an attempt to examine stories about NRTs as they were emerging in the late 1980s. Donna Haraway's contention that Western science should be seen as a set of stories and thereby identified as fully cultural (Haraway 1989: 3–8) was a main influence in this project. The decision to explore stories or narratives was an attempt to move away from definitive and simple statements about the meaning and significance of these technologies. This was in reaction to both uncritical celebrations of NRTs and to some feminist responses to them. In the 1980s, there had been considerable attention given to these technologies as the realization of male dreams of reproductive powers by both their advocates and their critics (for example Arditti *et al.* 1984). In contrast, I turned my attention to narratives about women's dreams and aspirations pertaining to these technologies. This reoriented the investigation, but it also complicated the picture. It forced me to acknowledge the ways in which these technologies were about *dreams* as well as *oppression*, and about *women's aspirations* as well as those of *male doctors and scientists*. But it was also my way of encouraging reflection about the dreams and aspirations of women (including feminists): how are they shaped? When and how might they be realized? When are they denied?

72 *Telling stories*

The analysis which follows in this chapter (and which is extended in Chapter 6) began as a talk which was given in a number of different forms and locations.[2] As such, there are deliberate rhetorical components which were used to draw out reflections about contemporary procreative stories (see also Franklin 1990, 1993; McNeil and Franklin 1993) and practices.[3] I was rather overwhelmed by the openness with which my audiences shared their tales of procreation and childlessness.[4] This analysis has been recast many times to take account of what I have learned from these exchanges and from the proliferation of writing on new reproductive technologies since the mid-1980s. Early in the twenty-first century, new procreation narratives are being written, spoken and lived continuously.

Originally, I wanted to work in what has been increasingly regarded as the dangerous terrain of 'women's experience' (see Riley 1988; Scott 1992). On the one hand, as I shall discuss more fully on p. 76, I was very aware that women had been pushed out of the picture in accounts of NRTs and, for this reason, focusing on their stories seemed strategically important. Nevertheless, I emphasize *work* here as a way of registering the now developed critique of the attribution of touchstone status to 'women's experience'. Denis Riley (1988) and Joan Scott (1992) are two of a number of insightful feminist analysts who have demonstrated the dangers of essentializing or reifying women's experience in social or political adjudications. Mindful of these dangers, I took stories about women's reproductive aspirations as the *starting* point, not the *end*-point, for my analysis. Furthermore, all accounts of experience are mediated and this was particularly obvious in dealing with reproductive stories, which appeared with increasing frequency from the mid-1980s onwards in the mass media in the countries in which I lived (Canada, the United Kingdom and the United States). This background fuelled my efforts to open up the category of 'women's experience' as it had been used with reference to NRTs.

Monolithic invocations of 'women's experience' can deny or obscure diversity and specificity, and, as I shall argue, this obscuring has been an important feature of the public discursive life of NRTs. Because of this and despite some reservations, I did want to signal specific aspects of my own background in undertaking research on issues of procreation. I was wary of analyses that loom as 'news from nowhere'.[5] Hence, it seemed important to identify myself as a white, now middle-class (formerly working-class), Western, heterosexual woman who was childless for social reasons. As my initial research in this field coincided with a sometimes difficult process of confronting and confirming my childless state, this also informed my approach to these issues. However, as a privileged, white, Western, heterosexual woman, and given my dissatisfaction with the narcissistic orientation of some recent feminist research, I did not want this to be the main focus of the piece. Indeed, I wanted to encourage reflexivity about the cultural values generated in new reproductive practices and about their implication in the generation of diversity – with reference to the allocation of attention and resources to *particular* women's experiences and dreams in the contemporary world.

Elspeth Probyn's *Sexing the Self* (1993) develops and amplifies the interest in politically strategic autobiographical work with which I have struggled in this chapter. Like Probyn, my intention is to draw attention to the constructed nature of autobiographical accounts and to the variability of their meanings depending on context, yet to try to use this autobiographical work provocatively and politically. My deliberately rhetorical reconstructions of procreative stories are in the spirit of Probyn's (1993) emphasis on the need for anti-essentialist, autobiographical work that pushes out to 'the other'.

The core of this chapter is an exploration of a set of procreation stories as a way of investigating the meaning and significance of NRTs. I begin with an examination of the term itself. From there I consider various ways of telling the story of these NRTs, before looking in more detail at particular, women-oriented narratives that have emerged around procreation issues.

Coming to terms with NRTs

It is helpful to begin with the term NRTs itself. These have been defined as 'all forms of biomedical intervention and "help" a woman may encounter when she considers having – or not having – a child' (Arditti *et al.* 1984: 1) or, more simply, as 'those technologies which facilitate, manage or prevent reproduction' (Throsby 2004: 9). The term came into use during the 1980s amongst social researchers (especially feminists) studying developments in the field of reproductive science and medicine. The term does not, in itself, specify the orientation of the technology and thus it encompasses contraceptive, as well as pronatal conceptive and birthing technologies. Generally, NRTs emerged as a collective designator of the range of reproductive technologies that became available from the 1960s onwards and it became strongly associated with the rapid development of the field of reproductive medicine in the last decades of the twentieth century. The term NRTs also has had special resonance within critical feminist scholarship in this field, which was particularly vocal in the United Kingdom, North America and Australia during the 1980s and early 1990s (C. Thompson 2005: 55–68).

The term NRTs is, in fact, somewhat misleading. In the first place, not all of the technologies which fall under its rubric are 'new'. For example, donor insemination (DI) has a long history stretching back several decades prior to the so-called revolution in reproductive technologies of the late-twentieth and early twenty-first centuries (Throsby 2004: 11).[6] In this and other cases, perhaps it would be more accurate to say that there has been more extensive use and routinization of such techniques. Moreover, the term NRTs begs the question of *how* these techniques are new. Some feminists have suggested that they are not new, but rather part of a long-term pattern of social relations involving the intervention of mainly male medical experts in women's procreative processes (Duelli Klein 1989).

This brings me to the adjective 'reproductive' as it is used in this context. Ruth Hubbard (1990) insists that this is a misnomer because we *do not reproduce ourselves*. In this sense, the term 'procreative' may be preferable.[7] Nevertheless, some commentators contend that developments in this field suggest that a desire for reproduction in a broad social, if not specifically biological, sense does fuel the investment in this technology (see O'Brien 1981; Noble 1992: 279–86). This is one of the reasons that I have chosen to use NRTs to label the developments with which I am concerned in this chapter.

In *Making Parents* Charis (Cussins) Thompson notes:

> A decade or more ago, the technologies were collectively referred to as 'artificial reproductive technologies,' and expressions like 'artificial insemination' were standard within the field as a whole. The word *artificial* has fallen gradually out of use, and the *a* in some well-known acronyms like (ART) has come to stand for *assisted*, not *artificial*.
> (C. Thompson 2005: 140, italics in original)

Indeed, the term 'assisted reproduction' has become common currency and some researchers prefer this more specific term to the broader rubric of NRTs. Some feminists have followed this pattern, especially when their main focus is on pro-natal technoscientific developments (see, especially, C. Thompson 2005). However, there are normative connotations of this label which deter me from using it. Although the term ARTs (assisted reproductive technologies) acknowledges that medicine and science are not the *only* agents in this process (since they 'assist'), the absence of direct specification of other agents (particularly the women who are the main recipients of such treatment) and its connotations of benevolence enhance the positive image of new reproductive technologies. In a study of media representations of new reproductive technologies, Sarah Franklin argues that 'the need for scientific assistance to human reproduction' (Franklin 1993: 528) is now widely and insistently highlighted. With mounting concern about infertility in many Western countries and media portrayals of conception as a remarkable achievement, as Franklin explains, 'the necessity for technological assistance thus comes to be seen as a product of nature itself' (Franklin 1993: 540). Marilyn Strathern contends that 'as long as some elements of the entire process of childbirth can be claimed as "natural", technological intervention appears enabling' (Strathern 1992b: 56). She demonstrates the ironic and irrepressible quality of this labelling when she observes that the framing of these technologies renders them unimpeachable, since 'they cannot "fail" to assist, even where the desire itself is not the end realised' (Strathern 1992b: 58). She also notes 'the new language of gifting that accompanies the propagation of late-twentieth-century technologies' (Strathern 1992a: 207). A further disturbing dimension of this term is that it seems to *presuppose* a positive reproductive outcome – that these technologies will result in childbirth. This obscures their pattern of failure and the often costly unsuccessful investments in and engagements with these technologies (particularly by women), which have been a crucial part of the experience of them in recent decades (Throsby 2004).

Karen Throsby suggests that 'the term NRTs remains useful because it recognizes these technologies as productive of, and products of, an intersection of ideologies of science, technology, reproduction and gendered bodies at a particular historical and cultural moment' (Throsby 2004: 10). I endorse this appraisal and, like Throsby, I continue to find this term useful, although I am mindful of the imprecision of the designation 'new'.[8] For this reason, and because of my concern about the problematic connotations of the replacement designation – ARTs – already discussed, I use the term NRTs throughout this chapter. The use of the term NRTs also signals my affiliation with a tradition of critical and watchful feminist scholarship and activism around developments in reproductive technoscience and medicine.[9]

Another source of confusion about the terms in this field results from the fact that some of the practices with which these terms have been associated do not necessarily involve high technology or any technology at all. Donor insemination is a case in point, realizable as it is in many 'low-tech' or, as some would claim, 'alternative-technology' modes. Likewise, surrogacy is often linked to NRTs (Spallone 1992: 208). Nevertheless, not all forms of surrogacy involve technology. It is often identified with technology because it can be combined with many of the practices listed previously, including IVF. Moreover, it is a practice which, like those associated with NRTs, seems to disturb apparently established patterns of reproduction and which raises some similar legal and ethical questions.

However, the bunching of 'low-tech' or virtually 'no-tech' reproductive practices and surrogacy with NRTs also indicates the strong inclination to make biomedical experts the custodians of all 'artificial' modes of reproduction. In Britain, at least, the non-technical nature of these practices is seldom observed. Surrogacy, donor insemination and the then available range of NRTs all came under the scrutiny of the Warnock Committee, which was established in 1982 and reported in 1984 on appropriate regulatory policies to the British government (Committee of Inquiry into Human Fertilisation and Embryology 1984). The difficulty (indeed, impossibility) of maintaining surrogacy under the regulation of biomedical experts was probably one of the reasons that it was the only form of 'human assisted reproduction' that the Warnock Committee did not recommend should be permitted; although there was a minority report advocating its legalization (Committee of Inquiry into Human Fertilisation and Embryology 1984).[10]

The range of technologies that might figure under the rubric of NRTs is extensive. The list of conceptive technologies could include: donor insemination, in vitro fertilization (IVF), amniocentesis, embryo transfer and freezing, ultrasonography, sex pre-selection, gamete intrafallopian transfer (GIFT), chorionic villus sampling, laparoscopy, zygote intrafallopian transfer (ZIFT), tubal embryo transfer (TET), intracytoplasmic sperm injection (ICSI), ooplasm transfer and pre-implantation genetic diagnosis (PGD). Moreover, this is by no means an exhaustive list.

While this list is extensive and has been continually growing since the late 1970s, IVF has been the main focus for much of this period. It was through IVF that the birth of the first successful 'test-tube baby' (Louise Brown) was realized in 1978. As Karen Throsby (2004) has argued, IVF has remained the 'core technology' in the proliferation of pro-natal, conceptive biotechnologies from 1978 into the first decade of the twenty-first century. It has also been the main focus of popular storytelling about reproductive technology during this period and for this chapter.

Different ways of telling the NRTs' story

There are many ways of telling the story of new reproductive technologies. For myself and a number of feminist researchers and campaigners of the late 1980s and early 1990s there was a strong inclination to tell the story of these technologies as part of the evolution of male-dominated, professional, medical and scientific regulation and control of women's reproductive functions – continuous from its beginnings with forceps and the discrediting of midwives.[11] This version of the story could be illustrated in a number of ways. A case in point can be found in *The Third Report of the Voluntary Licensing Authority* (VLA) (1988),[12] which referred to the need 'for a more precise definition of pregnancy' for use in measuring the success rate of IVF. The definition that this Licensing Authority proposed was 'the presence of the fetal heart on ultrasound' (Voluntary Licensing Authority 1988: 18). This definition of pregnancy emphasized the information provided by technology and reliance on the mediation of experts with no reference to the women involved.[13] The patriarchal dimensions of reproductive science and medicine have been traced by a number of feminist researchers. Some have identified 'a continuous and teleological process wherein patriarchal medicine monopolizes control over women's procreative bodies and reduces women to passive objects of medical surveillance and management' (Sawicki 1991: 76; see also Balsamo 1999: 93).

However, during the late 1980s and into the 1990s there began to be more investigation of the complex patterns of women's agency within the general picture of patriarchal science and medicine (Treichler 1990; Sawicki 1991: 67–94). Judith Waltzes Leavitt's (1986) history of birthing in the USA from 1750 to 1950 and the examination of women's resistances to medically constructed models of female bodily functions undertaken by Emily Martin ([1987] 1992) were in this mode. In addition to highlighting women's agency in general, these researchers paid particular attention to class and sometimes ethnic differences in analysing women's encounters with reproductive science and medicine. This was related to a wider feminist questioning of 'top-down' models of power and control (Sawicki 1991: 82; see also Arney and Neill 1982), of the class and ethnic specificity of some conceptions of control (Martin [1987] 1992), of the view that there is an '"authentic" [female] body waiting to be released from the bounds of medicine' (Lupton [1994] 1995: 160) and of uniformity and conspiracy amongst scientists and physicians (Treichler 1990: 118).

Feminist research on NRTs was been influenced by and has itself constituted part of the burgeoning of science studies since the 1990s. The changing climate for feminist activism and the poststructuralist turn in social research also influenced research in this field (see C. Thompson 2005: chs 2 and 3). Against this background, Charis (Cussins) Thompson's concept of 'ontological choreography' (Cussins 1996; C. Thompson 2005: esp. 8–11) became something of a marker of a new mode of feminist analytical research around these technologies. Deriving it largely from her ethnographic work, Thompson uses the term 'ontological choreography' to denote 'the dynamic coordination of the technical, scientific, kinship, gender, emotional, legal, political, and financial aspects of ART clinics' (C. Thompson 2005: 8). She traces the complex patterns that, when successful, lead 'to new kinds of reproduction and new ways of making parents ... ontological innovation' (C. Thompson 2005: 9). In contrast, Karen Throsby (2004) oriented her research not around the *procedures* of the clinic but, rather, around the *processes* recounted by women who have undergone IVF treatment unsuccessfully. She traces the material and discursive work oriented predominantly around normalization which constitutes their IVF experience. The sophisticated research of Throsby and Thompson provides distinct feminist twenty-first-century stories about NRTs. In fact, some feminists have started to assemble their stories about feminist responses to NRTs – delineating and demarcating different phases in this tale (McNeil 2001; C. Thompson 2005: esp. ch. 2). I shall consider these feminist meta-stories, as they might be called, later in this chapter.

In addition to accounts of patriarchal medical science, the story of NRTs must also be told as a tale of corporate, high-finance medicine and science. As Charis Thompson has observed:

> The companies that were making fertility drugs underwrote the clinical expansion of assisted reproduction technologies more or less since the birth of Louise Brown, with Bourne Hall (Steptoe and Edward's ART facility in the United Kingdom, which was set up after the birth of Louise Brown) famously being funded by Ares Serono.
> (C. Thompson 2005: 233)[14]

Pharmaceutical and biotechnical companies and private clinics (Pfeffer 1992, 1993: esp. ch. 6) are key institutions in this version of the story, as are insurance companies and law firms, particularly in the USA. It would be a formidable task to fully map the political economy of these developments. Nevertheless, there has been more attention given to the bio-economy of NRTs in recent years, and the research of Charis Thompson (2005: esp. chs7 and 8) and that of Catherine Waldby and Robert Mitchell (2006) has been important in this respect.[15] This has been accompanied by the acknowledgement and tracing of the consumer practices which have been generated in and through NRTs (see Throsby 2004: ch. 4; Fletcher 2006).

Such mapping would also have to register national differences which have been important in the development and use of NRTs (see Gunning 1990; Jasanoff 2005: ch. 6). Indeed, the NRTs' tale can be narrated as a set of stories about the making of national differences. For example, there are interesting contrasts between patterns of development of NRTs in Britain and the United States (Pfeffer 1992; Jasanoff 2005).[16] One crucial theme in the British story concerns the way apprehensions about these technologies have focused on the issue of commercialization. Hence, during the 1980s particularly there was a widespread feeling that keeping commercial agencies under control and marginalized was a key issue. The Warnock Committee (1984) fell into line with this orientation and ignored the complex political economy (including pharmaceutical companies' involvement with clinics and the links between the National Heath Service (NHS) and private clinics) which already underpinned NRTs at the time of their deliberations.[17] Despite the efforts to minimize corporate involvement, as Naomi Pfeffer (1993) has shown, during the 1980s NRTs were mainly provided through private services in Britain and this continues to be the pattern in the early twenty-first century. By contrast, in the United States during the 1980s there was pressure to keep the state at a distance from this field and to leave it to the market, thereby avoiding the imperative for government regulation (Powland 1988: 205). The suspicion about government interference has resulted in a complex state-specific pattern of provision, with privacy being the watchword in the United States, as C. Thompson (2005) has recently shown. Marion Brown, Kay Fielden and Jocelynne A. Scutt (1990) have traced another national pattern and argued that up to the early 1990s the development of NRTs in Australia was shaped by the specific features of that country's economic and industrial development. Common-law traditions in English-speaking countries (Duden 1993a: 101) and the powerful controversy generated by the abortion issue in the USA (Samuels 1995: 175; Duden 1993a; C. Thompson 2005) are two further factors in national contexts which have influenced the development of and responses to NRTs. The most elaborate national evaluation of NRTs was undertaken by the Canadian Royal Commission on New Reproductive Technologies, which issued its report in 1993. This report offered a distinctively Canadian, if contested, perspective on these technologies (Royal Commission on New Reproductive Technologies 1993).[18]

A final and overlapping narrative which weaves into the various national and international stories about NRTs would be an account of the development of professional science and medicine in this field. In the early 1990s, some feminist research did begin to explore the history of experimentation and the emergence of professional competition in this field (see Crowe 1990b; Koval and Scutt 1990). There is certainly a story to be told about the ascendancy of reproductive medicine, which was very much a Cinderella specialism in Britain and elsewhere until the last quarter of the twentieth century. An account of the making of careers and, in Britain, of the making of heroes of reproductive science and medicine would be both intriguing and valuable (C. Thompson 2005: ch. 7).

Despite my interest in the preceding ways of telling the story of NRTs, my focus here is somewhat different. Although I have been influenced by and shall draw on the kinds of accounts I have sketched, I will not be looking at these grand narratives. Instead, this chapter is organized around three different women-oriented narratives which pertain to women's dreams about NRTs. These are:

1 recent personal, but publicly available, narratives of women about their relationship to procreation;
2 a feminist utopian vision of reproductive technology as liberating;
3 the clichéd but everyday story that attaches to much domestic technology – about NRTs as labour-saving.

These are narratives about women's dreams and expectations pertaining to procreation and reproductive technology.[19] The focus on women's voices is strategic: to contest the silencing of women (Haraway 1992: 311–12; Taylor 1993: 609), the pushing of them out of the picture (Petchesky 1987; Spallone 1992: 207; Stabile 1994: 68–98), what Carol A. Stabile has labelled 'the visual and symbolic exclusion of women' (Stabile 1994: 89) in foetal-centred representations of reproduction. Moreover, the exploration of these narratives provides rich routes in exploring the relationship between women's desires and needs and the development of NRTs.

NRTs dream narrative 1

True reproductive story: version 1

I shall begin with a story which has become familiar in the Western industrial world during since the last decades of the twentieth century:

> I can't have a baby. Like many of the heterosexual, middle-class, white women of my age group in the Western industrial world, I moved through my twenties concerned with contraception and avoiding pregnancy. Financial independence and birth control were to be my tickets to freedom and I invested in both early and completely. As I moved into my thirties, I began to think about having children. The past ten years or so have brought me to a gradual and sometimes painful confrontation with the fact that I can't have children. As time slips away, it seems increasingly unlikely that I'll ever become a mother.

This kind of narrative has become familiar in representations of reproductive technologies. In autobiographical form or in its third-person version (e.g. 'Jane Smith can't or couldn't have a baby') it now constitutes one of the distinctive genres of our time and our society. The mass media in Western industrial countries like Britain and the United States have used such accounts as the way into reports about NRTs since the 1980s. Perhaps no other genre (except possibly the narratives of people with AIDS/HIV) typifies introductions to the challenges associated with late twentieth- and early twenty-first-century high-technology science and medicine.

80 *Telling stories*

We have become so accustomed to these stories that we expect a quick medicalization of the story: enter doctors and scientists. A published account illustrates this familiar trajectory:

> At thirty, securely employed as a lecturer, the obligatory *Hitchhiker's Guide to Europe* behind me, I found myself reconsidering 'motherhood'. ... I 'came off' the pill to become a mother and climbed down into pain. ... G recommended a private gynaecologist who would prescribe Clomid ... to be combined with temperature charts of daily readings taken 'before you put a foot out of bed'.
> (Maggie Humm, quoted in Duelli Klein 1989: 36–7)

The trajectory in this and many similar accounts leads to the elaborate details of an immersion in IVF treatment: a process which, in its most basic form, involves chemical stimulation of ovulation to produce several eggs, retrieval of these eggs from a woman's ovary, mixing them with sperm in a culture medium, outside her body, in a laboratory, and placing some of them back into her womb. This is the standard technical account of basic IVF.[20] It is an inadequate, stripped description of the process and of the technology which minimizes the extent of and the investment in both.[21] Indeed, Karen Throsby has shown that it is important to distinguish between the '*procedures*' (technical) and the '*processes*' of IVF (how women and, in a rather different way, their partners *live* their treatment) and she observes that the former have garnered more attention than the latter (Throsby 2004: 12).

Narratives like the one just cited began to appear in the media in Britain and in some other Western countries in the mid-1980s and they became moral tales of our high-technology societies in the last decades of the twentieth century. For the most part, these stories are structured around the assumption that it is perfectly reasonable that such investment in NRTs should be made.[22] The readers or audience are encouraged to identify with the narrator's goal of a happy ending in technological triumph and feminine fulfilment through motherhood, or, at least, a joyous birth. Indeed, the latter is usually the end-point of such narratives and readers/viewers are encouraged to read off the former (feminine fulfilment) from the latter (a joyous birth) (Franklin 1990).

True reproductive story: version 2

Returning to the opening true story and putting it in a different context, I can present it not as a generic form, but as *my own personal* story:

> I can't have a baby. Like many of the heterosexual, middle-class, white women of my age group in the Western industrial world, I moved through my twenties concerned with contraception and avoiding pregnancy. Financial independence and birth control were to be my tickets to freedom and I invested in both early and completely. As I moved into my thirties, I began

to think about having children. In my forties, I came to a gradual and sometimes painful confrontation with the fact that I couldn't have children. As time slipped away, it became increasingly unlikely that I would ever become a mother.

I can break some associations of such stories with high technology by explaining that my painful confrontations were neither biological nor technological, but social and political. It was these things in the form of the unwillingness of male partners to share in the responsibilities of childrearing with me, the lack of provision for childcare in the countries in which I had lived, the pressures on full-time professional working women, the isolation of single mothers, the conventions of coupledom and other social constraints that made me feel that I couldn't have a child.

The point about my personal story is not to evoke sympathy, but to underscore that in late twentieth- and early twenty-first-century Western culture it has been stories of biological difficulties and of the promise of reproductive technoscience and medicine that have preoccupied the media and captured the popular imagination.[23] This is illustrated by the comment in a rather progressive infertility handbook which proposed that: '[i]f a single woman has no fertility problem, other than the absence of a partner, she can overcome this herself either by self insemination or by finding a suitable partner' (Biggs 1989: 75).[24] In fact, I would suggest that since the 1980s British and North American cultures have tended to construe biological barriers as challenges, while the social and political imaginary for addressing other kinds of problems (particularly in the area of reproduction) has become impoverished. Indeed, in this realm these are societies which prioritize technical fixes and struggles. Hence, it is not uncommon to encounter empathy with middle-class women who embark on programmes of IVF, scorn for poor women who become pregnant 'unwisely' and ignorance or denial of the ways sexist societies inhibit some women from having children.

The so-called 'Virgin Birth' controversy in Britain in 1991 vividly illustrated the tensions which can result in a society in which preoccupation with technical fixes takes precedence over social or political negotiations. The controversy erupted when a press leak brought attention to the fact that women who were not in heterosexual relations were seeking and securing IVF treatment (Estling 1991; *New Scientist* 1991; *Nature* 1991). This was a fascinating episode not least because it highlighted doctors' positions as the main gatekeepers to women's reproductive rights in Britain.[25] However, it also showed that some women who were not in heterosexual relationships could try to become mothers and gain access to intimate relations with children through becoming medical patients. In short, being an IVF patient, they might 'pass' as potential socially sanctioned would-be mothers. Nevertheless, the furore against this practice in some quarters (for example in the *Daily Mail*) indicates the widespread expectation that these technologies should be used to bolster the traditional nuclear family and conventional heterosexual relations. At this time, this expectation was in tension

82 *Telling stories*

with the strong social investment in the medicalization of reproduction. In these terms, the established convention in early 1990s Britain that reproductive rights should be tied to heterosexuality came up against the assumption that these were matters of medical jurisdiction and doctors' regulation.[26]

Considered from a rather different perspective, this controversy showed single women (some of whom may have been lesbian) attempting, unsuccessfully as it turns out, to bypass social and political conventions and controversy. In seeking individual medical access to IVF in Britain at this time they could achieve part of what had previously been denied them and what direct social or political campaigns had not achieved: the possibility of becoming mothers. In this sense, they were opting for a technical fix which, at that moment at least, they hoped would allow them to sidestep direct confrontation with the social restrictions facing lesbian women.

True reproductive story: version 3

In both of the preceding versions of the 'I can't have a baby' narrative the tale was filtered through the prism of individual experience. However, the same story could be told from a rather different angle, as a somewhat detached social observer or social scientist – or as a visiting Martian anthropologist – might recount it:

> Towards the end of the twentieth and beginning of the twenty-first centuries, a group of relatively privileged women (mostly white and middle or upper class) in Western Europe, North America and Australia were not able to procreate. Considerable resources (financial and professional) were devoted to this problem in order to guarantee that these women could choose to have their own children.

Once again we are likely to associate this ostensibly more objective account of a social pattern with high-technology reproductive science and medicine. We might encourage our Martian anthropologist to regard this as a sign of our civilized, caring societies – as an instance of the devotion of financial resources and professional skills to meeting women's needs.

True reproductive story: version 4

This account can be given a further twist by placing it in a different context. It could be an observation about female academics in Britain and North America in this period. Although it has not been fully documented, there is good reason to believe that in Britain and the United States many such women remain childless and that they tend to have fewer children than either a comparable cohort from the general female population outside the academy or a cohort of their male colleagues.[27] There is no evidence that this pattern is due to genetic disturbances or biological incapacities amongst this group. It is true that this could, to some degree, reflect positive decisions amongst some of these women to remain 'child free'. This explanation notwithstanding, I would suggest that this may be part of a pattern of adaptation to social pressures and of survival strategies in sexist societies.

If I were to argue that considerable resources (financial and professional) should be devoted to this problem, many would be surprised and certainly not all would approve. Once again, this social pattern reveals a society which prioritizes biological and medicalizable problems, while shrinking away from more explicitly social and political problems. My hypothetical Martian might be shocked to discover that so many US citizens could be moved by the successes of NRTs and yet be so sanguine about the lack of a national programme for paid maternity leave and unfazed by high infant mortality rates amongst the poor in their cities.[28] I encountered what I considered to be an astounding denial of social patterns when I heard glib comments at the college where I taught in the USA in 1989–90 indicating that female academics teaching at that college had '*chosen*' not to have children when there was no maternity leave programme available. Related assessments have been made with reference to higher education institutions in Britain, Canada and elsewhere, although there is state maternity leave provision in Canada and the United Kingdom.

Generally in the last decades of the twentieth and the early twenty-first centuries in the sphere of biopolitics, Western societies such as Britain, Canada and the United States have become timid about tackling many social or political problems, while they relish technical struggle. In relation to NRTs this does suggest that the development of such technologies cannot be taken as *prima facie* evidence of a civilizing concern to facilitate women's desires for procreation. Indeed, the implementation of social changes (including good maternity leave policies, provision of high-quality, inexpensive childcare and other forms of maternal support) that would facilitate the realization of these procreative desires more universally remains strikingly limited in the United States, Britain and many other Western countries.

Instead, we could observe that these technologies have predominantly been *oriented* around *some* of the desires and needs of mainly white, middle- and upper-class, heterosexual women in the Western world. While there have been efforts to extend access in both Britain and North America which has brought some change in the demographic profile of users of IVF over the last two decades, white, middle- and upper-class women continue to dominate the picture (see Throsby 2004; C. Thompson 2005). Towards the end of her very positive overview of the evolution of ARTs (the term she uses) in the United States, Charis Thompson quotes the damning assessment of the executive director of a Native American Women's Health Education Resource Center who pronounced in 2004 that 'these technologies do not concern us because we do not have access to them' (C. Thompson 2005: 257).

Provisos about which women have access to these technologies have been an integral part of the administrative practices in this field of scientific medicine and I shall discuss the evolution of access in more detail in Chapter 6. In the

United States private clinics have had variable but generally limited state regulation, and, to a considerable degree, they have developed their own conventions about access. In Britain there have been higher levels of state regulation, initially through the VLA (established in 1985) and then, since 1991, through the Human Fertilisation and Embryo Authority (HFEA).[29] While lesbians' and single women's access to IVF and other related NRTs has increased in both countries over the last decade and a half, these women still do not have full legal rights to these technologies in Britain. Beyond this, financial constraints have determined the pattern of provision in a situation where public provision of NRTs has been restricted either by its uneven resourcing through the NHS in the United Kingdom or through differential state subsidy and the vagaries of private medical insurance coverage in the United States.[30]

Moreover, the conjured picture of technologies meeting the needs of women provides a rosy gloss on a much more complex and ambiguous picture of women's experiences of these technologies. Although the success rates of these technologies have improved over the last two decades, failure remains a significant feature of their employment. The Canadian Royal Commission on New Reproductive Technologies concluded in 1993 that '[a] substantial proportion of women undergoing IVF at fertility programs across the country are doing so when there is no evidence that, given their diagnosis or that of their partner, IVF will help them to conceive' (Canadian Royal Commission on New Reproductive Technologies 1993: I, 521). The situation has improved and the British HFEA indicates that current success rates (based on statistics from 2003–4) vary between 28.2 per cent for women under 35, and 10.6 per cent for women between 40 and 42 in UK clinics (www.hfea.gov.uk – Facts and Figures). However, as Karen Throsby (2004) fully documents, most IVF cycles still end in failure and this means that there are many more problematic and negative dimensions to the treatment than the widely circulated stories of joyful fulfilment convey. Meanwhile, as I have indicated, these technologies are being intensively developed and used while there is limited initiative in tackling many of the most widespread social problems associated with women having, rearing and nurturing children.

My final note on these narratives is that, while the versions used above are styled after widely circulated media stories that have become familiar in the Western world, they break with the latter in the presentation of the subject/narrator's position in relation to key social divisions (most notably class, ethnicity, geographical location and sexual orientation). Despite their highly personalized and often intimate forms, many media stories about new reproductive technologies project the voice of 'everywoman' – with little or no allusion to these crucial social divisions. This obscures patterns of social use and denies the significance of these divisions in the development of new reproductive technologies. Moreover, the presentation of women undergoing new reproductive technologies as 'everywoman' evokes considerable empathy and invites high levels of identification with their plight.

An older feminist NRT dream narrative

In the early days of what came to be labelled 'second-wave feminism' Shulamith Firestone reflected on how technology would fit into her vision of the potential feminist revolution. She claimed that it was 'woman's reproductive biology that accounted for her original and continued oppression' (Firestone [1970] 1979: 74). Firestone's was a technicist vision of that revolution, as she confidently predicted: 'Soon we shall have a complete understanding of the entire reproductive process in all its complexity'; and more prophetically: 'Artificial insemination and artificial inovulation are already a reality. Choice of sex of the foetus, test-tube fertilization ... are just around the corner' (Firestone 1979: 187). She bemoaned the fact that 'fears of new methods of reproduction' meant that the subject, 'outside of scientific circles, is still taboo' (Firestone 1979: 188). The lifting of this taboo and the expansion of technological modes of reproduction were seen as key elements in her feminist revolution.

Many of the technical features of Firestone's dream have become a reality, but they have not been realized as part of a feminist revolution. Of course, Firestone herself had no feminist perspective on the history of science, medicine or technology. The last three decades have witnessed a strong and multifaceted critique of what has been labelled patriarchal or androcentric science, technology and medicine (Easlea 1980a; Merchant 1980; Keller 1985; Harding 1986; Wajcman 2004). When viewed from the perspective of this body of research, NRTs would hardly be expected to be the media for the dreamed-of gender revolution. Instead, they would be placed within the now well-established and well-documented story of patriarchal professional expertise and technology. In short, for most feminists today there are not the expectations of liberating technical fixes of the sort that danced in Firestone's head.[31]

It is easy for those who have benefited from the work of feminists like Sandra Harding, Evelyn Fox Keller, Carolyn Merchant, Donna Haraway and countless others to be dismissive of Firestone's dreams and to feel superior in reflecting upon her naïveté. Nevertheless, it can be helpful to return to her analysis to delineate the old and new issues that have emerged around NRTs. It could be interesting to reconsider Firestone's arguments and explore *why* NRTs have *not* helped in the realization of the dream of female emancipation.

Here we could begin with the simple observation that in many accounts of the development of NRTs the term 'mother' is used in a manner which could be deemed inappropriate. One of the most established researchers in the field in Britain (Robert Edwards) referred to '*the mother*' being 'superovulated' (Spallone 1989: 79). Likewise, Robert Edwards, Patrick Steptoe and Jean Purdy, in one of the earliest articles on IVF, wrote: 'Human oocytes [eggs] had been taken from *the mother* before ovulation' (Edwards *et al.* 1970: 1307, quoted in Spallone 1989: 79, my emphasis). The latter comment appeared some eight years before *any* woman became pregnant through IVF, with the procedures being described. This substitution of '*mother*' for '*woman*' could be dismissed as a mere slip of the tongue or pen. Instead, I would suggest that the language of NRTs here and elsewhere is a sensitive indicator. In this case, it is indicative of a literal and substantive extension of women's maternal roles and responsibilities.

To explain this it is necessary to turn to the complex history of mothering and childcare in the Western world. There have been key moments in the elaboration of responsibilities associated with the job of mothering, particularly in specific forms for white, middle-class women in Western societies.[32] Freudian psychology, with its emphasis on early psychic formation, appears to be one key moment of this sort. Valerie Walkerdine and Helen Lucey (1989) have documented the emergence of the paradigm of the 'sensitive mother' in post-World War II Britain as a fairly recent, British phase in this history. In this respect, it could be argued that we are in the midst of a further revision of the job requirements of motherhood. This could be described as an extension of the biological expectations and temporal dimensions of the job based on a strong moral imperative. As I see it, the moral responsibility for motherhood is being pushed back further and intensified. NRTs are by no means the only element in this development, but they have been at the centre of this development.

NRTs involve considerable female responsibilities and body discipline in the hope of becoming pregnant. To use the language associated with Michel Foucault, these are new regimes of body regulation or 'disciplinary technologies' (Sawicki 1991: 83). This is obvious from the direct testimonies of women who have undergone IVF treatment who have spoken or written of their experiences (Duelli Klein 1989; Throsby 2004). Some of these are harrowing accounts in marked contrast to newspaper splashes about 'test-tube babies' and 'technological achievement'. Even when these reports indicate a smoother process and are relayed humorously, they indicate an elaborate and demanding process.[33] Continuous measurement of body temperature, regulated and carefully timed sexual intercourse, and frequent examinations by one or more medical experts are some of the elements in these regimes. Despite some variety in their form and duration, they cannot but impress upon a woman her responsibility for conception and birth. The result is an intensification of women's investment in procreation, realized in the regimented orientation and surveillance of her body for this purpose.

This is a complex process. When successful, this investment undoubtedly contributes to the euphoria that the media have been so eager to report. Nevertheless, often this itself has been overshadowed or even usurped by the attention given to the doctors or scientists involved. Moreover, much of this reproductive labour is privatized and hidden. Some women's own accounts have been published (Duelli Klein 1989; Modell 1989; Murdoch 1990). However, there remains a chronic under-representation of the demands of such treatment (Throsby 2004).[34] Many women are hesitant to discuss their experiences as they undergo treatment because of the intensification of pressure which may result from such disclosures. Indeed, as Karen Throsby (2004: esp. ch. 5) has shown, assuming responsibility for trying to manage the responses of those around them is, in itself, an important part of the work of IVF for women undergoing treatment.

The slips in the use of the term 'mother' by Edwards and other NRT scientists and clinicians do not seem insignificant in relation to the features of maternal responsibilities outlined here. With NRTs women are becoming oriented around motherhood – their lives being dominated by the possibility of motherhood – long before they give birth and often even if they never do. In this respect, NRTs can be viewed alongside other recent developments which have been pushing back the frontiers of maternal responsibility. For example, prenatal preparation has been extended and intensified. Moreover, in the United States since the 1980s there have been court cases about maternal responsibilities in relationship to the foetus. Drug taking or any activity which could be a potential threat to a foetus can become the basis of a custody case or other court interventions. The *New England Journal of Medicine* reported the case of a Wisconsin teenager who was held in secure detention for the sake of the foetus because 'she tended to be on the run' (Kolder *et al.* 1987: 1195; Spallone 1989: 44). In short, as Susan Faludi (1992: 459–66) and others have documented, in the United States women have been tried and found culpable if their behaviour appears not to be conducive to their obligations as mothers (Taylor 1993: 614–16).

It would seem that these developments may be associated with what Sawicki calls 'new norms of healthy and responsible motherhood' (Sawicki 1991: 84) or with what I would designate as an extension of maternal responsibilities. Certainly in Western countries we have become accustomed since the 1980s to more extensive preparations for pregnancy, including more attention to dietary regimes, consumption of vitamins and minerals, exercise and use of fertility drugs. In short, biological responsibilities are being pushed back long before conception or even before women are considering pregnancy. This was neatly exemplified in the popular film *Parenthood* (1989; directed by Rod Howard). The central characters are summoned to their son's school to discuss his difficulties. When confronted with the school principal's assessment that they have a problem child, the father (Gil Buckman, played by Steve Martin) turns to his wife (Karen Buckman, played by Mary Steenburger) and attributes this to the fact that she smoked dope when she was young. This charge is an amusing episode focused on maternal responsibility, but it is not an insignificant one.[35] It demonstrates the expectation that women are, or should be, in long-term preparation for motherhood: that maternal responsibilities begin early.[36] Another more serious manifestation of this came in the *US Public Health Service Experts Panel Report on Prenatal Care* published in October 1989, which advocated that, because so many pregnancies occur 'by accident', prenatal education should begin in high school.

This second public extension of material responsibilities is linked to NRTs in two ways. First, women have been charged if they fail to use such technologies. Thus, for example, legal action has been taken against women for not having Caesarean sections in the USA.[37] Moreover, the imperative to submit oneself to the range of available technologies in such circumstances is strong: amniocentesis, ultrasound, IVF and so on. Even if they are not actually legally prosecuted for not pursuing a full course of technological treatment, women can be made to feel and sometimes do feel guilty unless they do so,[38] although it is important not to underestimate their potential for resistance. My own discussions with women who have undertaken IVF confirms the reports of many feminist investigators that women often experience a 'technological imperative' – a strong feeling that they should avail themselves of whatever technological resource is available in their efforts to have a child. For example, repeated attempts to conceive with IVF are common because there is always what Christine Crowe describes as 'the lure of "next time"' (Crowe 1990a: 39).[39] Margarete Sandelowski highlights the 'never-enough quality' of conceptive technologies, observing that 'the techniques that constitute infertility therapeutics work in ways that compel their repetitive use' (Sandelowski 1993: 49). Almost all researchers in the NRT field comment that most women find it difficult to stop infertility treatment unless they become pregnant and give birth (Throsby 2004; C. Thompson 2005).

In addition, menopause, which was formerly considered to be the natural marker of the end of women's reproductive capacities, has recently itself become something of a challenge for some reproductive scientists and clinicians. As the stories of 'Britain's Oldest [62-year-old] Mum' which appeared in the UK in May and July 2006 indicate, the timeframe for pregnancy and motherhood has also been extended.[40] I shall consider the stories about postmenopausal pregnancies in Chapter 6.

Thus, in both the long and the short term, NRTs have been part of an extension of maternal responsibilities, investment and procreative activities. Moreover, as Irma van der Ploeg (2001) has demonstrated, women are increasingly required to undergo technological procedures as a means of dealing with male infertility and of facilitating foetal surgery. Hence, far from being an escape from the procreative process as Firestone predicted, NRTs have intensified its hold over women. These technologies have also been crucial in what Barbara Duden labels the recent 'epoch of fetal dominance' (Duden 1993a; see also Petchesky 1987; Samuels 1995; Berlant 1997: ch. 3). In her anthropological study of medical conceptions of menstruation, childbirth and menopause and of Baltimore women's responses to these, Emily Martin observed that the contemporary Western medical model makes pregnancy 'the end-point' for which women's bodies are seen to function. This is ironic because, as she points out, 'in our society the great majority of the time most women are not intending to get pregnant' (Martin 1992: 112). Likewise, Anne Balsamo commented towards the end of the twentieth century that in the United States 'even when not pregnant, the female body is also evaluated in terms of its physiological and moral status as a potential container for the embryo or fetus' (Balsamo 1999: 90). NRTs can be seen in this light – as not freeing women from the constraints of biological reproduction as Firestone dreamed, but rather intensifying its hold. In addition, the mass-circulated stories of NRTs create an image of women that is far different from either Firestone's or Martin's vision of women's reproductive freedom.

Reproductive daydream narrative

One feature of domestic technology which is striking is how insistent manufacturers and advertisers often are about the labour-saving potential of these products. For example, there used to be a range of washing machines in Britain which were called 'liberators' and the label 'convenience foods' has become commonplace. Technology targeted for domestic labour often comes with such labour-saving banners. Hence it is important to look quite closely at the forms of labour that specific technologies involve. This also brings us to our most quotidian technological dream narratives that revolve around hopes that any technology will eliminate or reduce labour.

Analysing NRTs in terms of women's labour indicates that they are not dream machines. In the first place, they do not eliminate women's work or effort in the reproductive process. IVF is the most vivid example of this, with the rounds of tests, record keeping, trips to the hospital or clinic mentioned previously. Given this, some women have to give up their paid employment to facilitate the process.[41] Karen Throsby's (2004) study traces the elaborate material and discursive work undertaken mainly by women undergoing IVF procedures. Even with surrogacy, which is very much associated with women getting others to do the work of carrying the baby and literally performing the labour, this pattern applies. In addition to the work of the carrying mother, few commissioning women turn to surrogacy without undergoing extensive tests and without having considerable medical treatment.

In the mid-1990s Carol A. Stabile detected a broad pattern of rendering the pregnant labourer invisible 'at this particular historical moment' (Stabile 1994: 94).[42] Since then displays of pregnant bodies have become much more commonplace in the UK media (Tyler 2001). Nevertheless, these have mainly been glossy and glamourized and there is still little attention given to the corporeal dimensions of pregnancy, let alone to the material-discursive work involved in IVF.[43] Moreover, with specific reference to NRTs, when pregnancy and birth are realized it is 'the achievements of medical experts' and the 'wonders of the technology' which are generally celebrated. The very term 'test-tube babies' exemplifies this. The mother of the first such baby in Britain, Leslie Brown, underwent considerable testing, travelled hundreds of miles to an unfamiliar city to attend Dr Steptoe's clinic, returned ill and in considerable pain, and carried her foetus to give birth to a daughter who was labelled 'a test-tube baby' and 'an achievement of modern science'.[44]

It remains the case that the media present biomedical experts and NRTs in a way which obscures or denies female labour. Moreover, such labour is often hidden by its very nature since, like most domestic labour, it occurs in the private domain. In the case of NRTs, its intensely personal form renders it invisible, sometimes even to the women who do it. Added to this is the fact that many women keep their infertility (possible or confirmed) to themselves because they fear the social stigma attached to their childless state. Others, such as the lecturer I quoted previously (Duelli Klein 1989), feel that if they 'go public' they will be subjected to yet more pressure.

90 *Telling stories*

The 'I can't have a baby' stories considered in this chapter are moral tales. To the generations of mainly white, middle-class women in the Western industrial world who have had access to relatively effective and available contraception since the late 1960s and who sought greater control over if and when they had children, they can serve as something of a reprimand. They play into the widely cultivated image of such women as selfish, self-seeking careerists. Here I am reminded of a cartoon in the Pop Art style with the caption 'I can't believe it. I forgot to have children!' When these stories are juxtaposed with the scare stories associated with NRTs of 'women leaving it too late', the moral becomes clear. This has been spelled out in medical journals such as the *New England Journal of Medicine*, which point the finger at feminism and careerism as causes of infertility (Fédération des Centres d'Etude et de Conservation du Sperme Humain *et al.* 1982; Faludi 1992: 48).[45] What is denied here is the fact that many women of the 'baby-boomer generation' (born between the late 1940s and the 1960s) in Britain and North America delayed having children and invested in the world of paid employment because we recognized that, in our sexist societies, remaining childless for some period was a prerequisite for any possibility of independence. It was not selfishness, but our awareness (sometimes articulated, sometimes not) of sexism which motivated (and still motivates) many women to avoid pregnancy in their twenties. I cannot help but think of the recent blaming of these women for problems associated with infertility and childlessness as the ultimate revenge on women who, like Firestone, optimistically set out to break the chains of reproductive biology.

Possibly even more insidiously, the 'I can't have a baby' stories are being used to reorient subsequent generations of women. The warning to them is 'You shouldn't put childbirth off!' 'You too might get caught!' 'Beware, the very devices you used to increase your control (the pill and other contraceptives) might mean you won't have the choice of having a child.' The advice literature on pregnancy increasingly advocates early medicalization to avoid this fate (Fédération des Centres d'Etude et de Conservation du Sperme Humain *et al.* 1982). Clinicians and scientists who have established their reputation in the development and use of NRTs have published warnings about women delaying pregnancy in medical journals (see Bewley *et al.* 2005). Their assessments have been relayed in newspapers and other popular media as health warnings for 'women who delay babies until late 30s' (Meikle 2005). They call out to women: 'Start early.' 'Get your body in shape now and think of everything you do in terms of its potential effect on your future childbearing capacity.' In short, even women not using NRTs are being recruited through stories about them to orient their lives and their bodies towards reproduction. In that process, feminism (its impact and significance) is often either denied or undermined.

Conclusion

This chapter has considered the terms and the stories through which the meanings of NRTs have been accrued and identified in the late twentieth and early twenty-first centuries. It began with a review of the changes in the designating terms in this field of technoscience and of some of the grand narratives that have told the NRT story. However, the chapter moved away from these grand narratives and was mainly oriented towards women's dream narratives associated with NRTs at different levels: from personal narratives, including my own about being an independent woman and having procreative choices, through Shulamith Firestone's ambitious dream of a feminist revolution through reproductive technology, to the more mundane and sometimes clichéd stories about domestic technology being labour saving. These explorations of dream narratives suggest that, although there has been some research in this field (Rothman 1986; Duelli Klein 1989; Throsby 2004; C. Thompson 2005), there should be further explorations of women-centred approaches to these technologies.

It has also been important to consider the various ways in which women's dreams do become focused on these technologies. For some women who are successful in giving birth with the assistance of these technologies, NRTs do help them realize some of their dreams. However, it is important to acknowledge that failure rates remain high in this field (Spallone 1992: 211–12; Throsby 2004) and that this group still consists predominantly of the most privileged women in the world (Western, white, middle-class, heterosexual women).[46] In general, then, only some women's dreams are realized through these technologies. Against the background of their promise I have shown that they also involve the extension, intensification and elaboration of women's responsibilities for motherhood and they are linked to increased surveillance of mothers by the state and medical authorities, as well as intensified self-surveillance. This is in the context of women already assuming increased financial responsibilities for children (in Britain, at least), particularly as lone mothers.

My examination of women's stories regarding their dreams and expectations around NRTs complements the already well-documented evidence of the increasing medicalization of pregnancy and childbirth. Pre-pregnancy screening and pre-implantation genetic diagnosis can be seen as further extensions of these developments (see Throsby 2004: esp. ch. 8). This is perhaps part of a larger fascination with the challenges of inner space as the technological frontier of late twentieth- and early twenty-first-century technoscience, particularly given the crucial links between NRTs and genetic engineering (Spallone 1992; Throsby 2004: ch. 8).

This relates to my broader observation that my examination of narratives about NRTs indicates that Western European and North American societies have become increasingly conservative about tackling some social and political problems, while they are ambitious about conquering natural or technological barriers. Steve Platt (1991) makes a specific link between the increased policing of women's reproductive rights in the United States and cuts in the welfare state in the early 1990s. Pat Spallone (1992: 16) and Dorothy Nelkin and Susan Lindee (1995) deplore the recent fixation with 'the promise of genetic solutions to social problems' (Spallone 1992: 16) associated with NRTs. Certainly it is striking to point out the contradiction of the flourishing of NRTs, whilst infant mortality rates amongst inner-city poor in the United States remain shockingly high, when Britain has the highest rate of child poverty in Europe and while Western television stations broadcast pictures of starving children in many other parts of the world.

My investigation raises questions about how NRTs and the popular stories associated with them fit into other contemporary social patterns. These technologies would seem to be part of the individualization and reorientation of dreams around choices which the market provides.[47] This is very much the framing condition for these new technologies. Moreover, the emergence of these NRTs and the narratives associated with them, particularly in the mass media, have also been linked to denials of feminist strategies and achievements – the so-called 'post-feminist' discourse or what Faludi (1992) labelled the recent 'backlash against feminism'.

Throughout this chapter, I have been trying to situate stories about NRTs in a broad context to show that they *are* about women's dreams and yet about so much else as well. My exploration shows that the form and nature of women's dreams also merit attention. This involves raising, as Ann Snitow (1991) has suggested, some rather taboo questions about motherhood and its contemporary significance for Western women (including feminists).[48] This might also entail the exploration of why the dreams of many contemporary women are so intensely focused on motherhood and of what Marilyn Strathern has designated as the 'Euro-American' 'widely shared cultural assumption that persons desire children of their own' (Strathern 1992c: 156).[49]

Another part of the context of NRTs which has been insufficiently explored is the panic about fertility amongst mainly white, middle-class North Americans and some Europeans. The term 'infertility' itself is a slippery term and, as Karen Throsby (2004: 13) has suggested, self-help and guide books often do not define it. The term is often linked to the specification of a time period of 'regular unprotected intercourse' which does not result in pregnancy. It is virtually impossible to assess levels of infertility, particularly when even this rather imprecise designation of infertility has changed noticeably in recent years. Susan Faludi explained that, while from the early 1990s a couple would be labelled as infertile in the United States if the female partner fails to conceive after one year of 'regular unprotected intercourse' (Faludi 1992: 47), this label was formerly applied only after five years. In Britain, the HFEA advises couples to seek medical advice if they have not conceived after two years of regular unprotected intercourse. Given that the market for NRTs is drawn from the pool of those who identify as infertile, NRTs should be seen, in these terms, as both creating and defining the problem, as well as potentially offering a set of possible solutions to it. Moreover, as this chapter indicates, the stories of NRTs themselves beget fears of infertility.

This apparent crisis around infertility in many Western countries may also be about social and political, as well as biological, reproduction – about hegemony. In any event, it is apparent that in entering into the dreams of mainly white, middle-class Western women who reproduce with the aid of Western technoscience, we are participating in narratives that are about Western scientific progress and reproduction. It may not be too farfetched to suggest that, amongst other things, these stories reanimate dreams of cultural, as well as biological, reproduction.

Overall, it is clear that, whatever NRTs may mean for individual women, including those for whom they may bring the possibility of biological motherhood, they are *not* the path to women's freedom on a grand or even on a mundane, day-to-day level. Nevertheless, those feminist dreams of greater freedom do live on, although regrettably they are less publicly discussed in the early twenty-first century. These are to be fought for through social and political change, not through technological fixes.

This chapter has traced the emergence of stories around NRTs from the last decades of the twentieth century in the UK and North America. Chapter 6 extends and develops the analysis of popular reproductive narratives and tracks the evolution of the 'I can't have a baby' narratives which have become closely associated with the development of NRTs.

6 Telling tales of reproduction and technoscience

> Making babies is not rocket science, though doctors would like us to think so.
> (Winterson 2001)

As Chapter 5 indicates, I have shared Jeanettte Winterson's concern with the making of babies in the late twentieth and early twenty-first centuries. Our common interest also extends to the development of new reproductive technologies during this period. More specifically, I have been intrigued by the stories told about these technologies and the attention they have garnered in the mass media in many Western countries during this period. This chapter pursues that interest from where I left off in Chapter 5 through an investigation of the development of mass media stories around and about this technoscience that have emerged in many Western countries since the mid-1980s.

Chapter 5 shows how I set out to examine the popular personal narratives associated with the emergence of the so-called new reproductive technologies in the late 1980s (McNeil 1993). As I explained, this was my way of trying to get some purchase both on what was happening in and around these technologies and on the then burgeoning feminist debates concerning them. The pivot of my analysis was the stark, deeply personal, but highly formulaic 'I can't have a baby' stories which had been linked to these technologies, particularly through mass media representations in the countries in which I lived during the late 1980s and 1990s – Britain, Canada and the United States. Below is the synthesized generic version of these that I presented in Chapter 5:

> I can't have a baby. Like many of the heterosexual, middle-class (in origin or in positioning), white women of my age group in the Western industrial world, I moved through my twenties concerned with contraception and avoiding pregnancy. Financial independence and birth control were to be my tickets to freedom and I invested in both early and completely. As I moved into my thirties, I began to think about having children. The past ten years or so have brought me to a gradual and sometimes painful confrontation with the fact that I can't have children. As time slips away, it seems increasingly unlikely that I'll ever become a mother.

As I explained, in a number of talks and an article (McNeil 1993) I played rhetorically with various versions of the 'I can't have a baby' story to explore the significance and resonances of its various forms and manifestations. My rhetorical experimentation was designed to investigate how these narratives worked and to encourage reflection about contemporary procreative stories (see also Franklin 1990, 1993; McNeil and Franklin 1993). The synthesized narrative used here exemplifies my rhetorical experimentation. Styled after widely circulated media stories that have become familiar in the Western world since the mid-1980s, it differed from them through the explicit presentation of the subject/narrator's position in relation to key social divisions and locators (class, 'race' or ethnicity, geographical location and sexual orientation.). As I noted previously, despite their highly personalized and often intimate forms, most media stories of new reproductive technologies projected the voice of 'everywoman' – with little or no allusion to these crucial social divisions and markers. This has obscured social patterns around access to these technologies and denied the significance of these divisions in their development. Moreover, the presentation of women speaking of their reproductive difficulties and aspirations as 'everywomen' invited empathy and identification. This was a crucial feature of this popular Western testimonial form during the last decades of the twentieth century.

In the late 1980s and early 1990s, I argued that such stories (in either first- or third-person or in a couple's versions) constituted a distinctive kind of narrative of the Western world at that time (McNeil 1993). I suggested that perhaps no other genre of narrative (except possibly, far more ambiguously, that of people with AIDS/HIV) served as an introduction to the challenges associated with high-technology science and medicine. Like the narratives of people with AIDS, these were moral tales of the high-technology, late twentieth-century Western world. They were stories of a new kind of technological prowess comparable to that associated with rocket science during the peak of the space race.

Fascinated by the apparent predictability and power of these stories, I used rhetorical moves in talks and in a written version (McNeil 1993) to explore their features. These narratives were predictable in at least two respects. First, the form of these accounts was standardized, whatever their specificities. Each account included (sometimes after the provision of a brief introduction or background) confrontation with difficulty in reproducing (the problem), some exposition of the suffering caused thereby, the solution (generally through scientific medicine), followed by happy resolution (usually the birth of a child). Despite their ostensibly personal nature, these narratives were highly formulaic. Moreover, they were predictable in the further sense that such narratives became the ubiquitous framing devices for media representations (including news items) of new reproductive technologies in the 1980s and 1990s. For example, the media coverage of the appearance of the Report of the Canadian Royal Commission on New Reproductive Technologies, *Proceed with Care* (1993), the product of the most extensive national investigation on this topic ever conducted, exemplified this pattern. In October 1993, on the eve of the release of this report producing a typical specimen of the genre, the CBC (the Canadian

Broadcasting Corporation) radio news opened with the story of a couple who were apparently infertile and who were looking to NRTs to solve this problem. This story was the lead-in to the CBC account of this report.

There is much more that could be said about my earlier exploration but my interest here is with subsequent developments in and around these narratives. So I shall consider two examples of twenty-first-century reproductive narratives that appeared as a newspaper diary entry and a short newspaper magazine report.

Two early twenty-first-century tales of reproduction

> 'Trying: Leah Wild's IVF Diary': The final IVF diary entry sees the twins surrounded by people in white coats.
>
> It took 10 people to make my babies. The Professor of the Assisted Conception Unit, the Pre-implantation Genetic Diagnosis (PGD) coordinator, the embryologist, two cytogeneticists, the team of four doctors specialised in reproductive medicine, the consultant obstetrician, and my boyfriend who provided the basic material. Just as the NHS advertising campaign says, everybody's life depends upon more than a couple of responsible parents.
>
> (Wild 2001)

In 2000–1, the *Guardian*, one of Britain's most widely circulated 'quality' newspapers, carried occasional diary columns in its tabloid-size section (called *G2*). Intermittently, a set of writers provided first-person accounts of a variety of experiences, including farm life. Between June 2000 and March 2001, Leah Wild contributed to this series with her 'IVF Diary' ('Trying: Leah Wild's IVF diary', www.guardian.co.uk/parents/story/0,3605,334535,00.html). In irregular instalments she presented a sequential account of her difficulties around reproduction (related to a foetal genetic disorder she carried), the overcoming of this barrier to reproduction through technoscientific intervention (in the form of pre-implantation genetic diagnosis and in vitro fertilization), pregnancy and birth.

Wild's 'IVF Diary' stretched and segmented the standard 'I can't have a baby' narrative considered in Chapter 5 in accordance with the requirements of the *Guardian* occasional diary format, as she provided a 'blow-by-blow' account of her reproductive problems, medical treatment, pregnancy and birth. The last entry, which is my main focus here, provided the resolution and ending of the story. The opening sentence of this article starkly registers the complexity of early twenty-first-century high-technology 'assisted reproduction': 'It took 10 people to make my babies'. Wild's narrative is inflected with an apparently progressive attitude to social change, which was likely to appeal to the *Guardian*'s liberal-left readership. Her insistence that children are social products and collective responsibilities and hints that the traditional nuclear family may be an outmoded institution in the contemporary British setting provide her leitmotif:

> My boyfriend and I are the babies' parents. . . . We also happen to be their biological parents. Although we'd still be their mum and dad even if we

weren't. But our babies have another set of adults responsible for producing them – the PGD medical team.

(Wild 2001)

Although this diary ends, in accordance with the convention of such reproductive narratives, in the euphoria surrounding the joyful birth of, in this case, two healthy babies, Wild does signal that the story does not end here:

It took 10 people to make my children. It will take far more than that to raise them.

(Wild 2001)

Despite its radical gloss, its gestural resistance against conventional closure and its production in instalments, with attention to detail, Wild's diary conforms to the pattern of the established 'I can't have a baby' narratives. Moreover, it extends and intensifies the conventional celebration of the achievements of technoscience characteristic of such stories. There is a fulsome acknowledgement of the scientific-medical team and of the technological procedure employed in the reproduction. Indeed, she describes her children as 'the *product* of PGD' (my emphasis). In fact, Wild praises the medical-scientific team, not only as '*scientific*', but also as '*social* pioneers', implying that they are in the vanguard in breaking the conservative stranglehold of the traditional nuclear family. Leah also acknowledges the contribution of her boyfriend – the genetic father – in the production of the twins. Nevertheless, a notable feature of this narrative is that, despite its autobiographical form, the account effectively erases Wild's own contribution to the production of her twins. This denial of her own agency is in tension with the autobiographical form. In this respect, Wild's narrative recapitulates the erasure of women in representations of reproduction which, as noted on p. 76, feminists had criticized in the 1980s and early 1990s (Petchesky 1987; Spallone 1992: 207; Stabile 1994: 68–98).

'Amanda's story'

Amanda Pearson, 34, and her husband, Colin, from Ashford, Kent, endured eight years of repeated miscarriages. Finally, they discovered that Amanda had a genetic translocation, where part of one chromosome becomes attached to another, and that pre-implantation genetic diagnosis was her only hope. Amanda and Colin are now the delighted parents of eight-month-old Joshua.

(K. Williams 2001: 39–40)

Whereas 'Leah Wild's Diary' extends and inflects the 'classic' 'I can't have a baby' narrative form, 'Amanda's story' provides, in its opening three sentences, a condensed representation of this genre. Appearing under this simple, personalized title, this narrative literally frames an exposition of recent developments around genetic testing in the supplement of the *Mail on Sunday* – *You* magazine.

98 *Telling stories*

In the opening three sentences all the elements of this reproductive narrative are laid out: the confrontation with a problem around reproduction, the medical-scientific diagnosis and solution, and the resolution of the problem through a joyful birth. What follows, in the rest of the column, is the elaboration of the details of this story, including information about the medical diagnosis and treatment. This includes the representation of this couple as at the leading, testing edge of technological development. The article reports that, after initial diagnosis, they 'were told that PGD could help, but that the technology was still being developed. So we waited for two long years.' Reference is made to the cost and anxiety associated with the PGD procedures and to the couple's prospects for future reproduction, with their reported comment: 'If we can't have any more children, though, we will have to do the same again. But it would definitely be worth it.'

These two examples are journalistic exemplifications of the 'I can't have a baby' narratives discussed previously, with some distinguishing features. The first autobiographical account tells a reproductive story in instalments, the excerpt considered here being the final instalment. The second narrative appears as a framing 'human-interest', third-person overview (peppered with direct quotations from the couple) of their problems and ultimate success in reproducing, which accompanies a more factual exposition of developments in new reproductive technologies for a popular readership. Nevertheless, as I have indicated, these are both exemplars of the genre of reproductive narrative identified earlier. True to this form, these are highly personal stories – '*Leah Wild*'s diary' and '*Amanda*'s story' – without any specification of social status or markers (for example class or ethnicity), but which nevertheless claim a certain representativeness. Wild's is not just her personal reproductive diary, it is an '*IVF* diary', and Amanda's story frames and highlights the need for the technological innovation presented in the article as a whole.[1]

Assessing the significance of such narratives requires that they be situated in a broader historical and theoretical context. In the next section (pp. 98–102), I begin such assessment by locating these narratives in relation to research undertaken by Ken Plummer and Lauren Berlant.

Intimate citizenship and 'I can't have a baby' narratives

Western personal narratives about obstacles to reproduction can be related to the new modes of 'intimate citizenship' (Plummer 1995, 2003) and manifestations of the 'politics of intimacy' (Berlant 1997: 7) associated with the recent explosion of testimonial forms in public life in Britain and North America. This section of the chapter examines salient aspects of Ken Plummer's and Lauren Berlant's interpretations of these new forms of citizenship in the Western world and the popular narratives with which they are associated. I bring Plummer's and Berlant's work into dialogue around their critical perspectives on emerging forms of the politics of intimacy to generate a synergy for my own analysis. Their research is used as a theoretical prism through which to view the development of 'I can't have a baby' narratives over the last decades of the twentieth and the first decade of the twenty-first centuries. But I also use my own research to cast light on these new perspectives about recent forms of public life and their popular narratives. This section begins with an exposition of Plummer's and Berlant's analyses of recent manifestations of the politics of intimacy. An exploratory triangulation between the work of these two critics and my own investigation frames the subsequent account of the three defining features/processes that have characterized the pattern in the popular circulation of 'I can't have a baby' narratives during the past two decades or so: resilience, proliferation and differentiation.

My initial interest in recent reproductive narratives was primarily in their social and political significance. Retrospectively, and particularly through the prism of recent research on public intimate narratives and for other reasons I will consider, it has become easier to articulate their significance. Although such reproductive narratives are not necessarily *Sexual Stories* as designated in the title of Plummer's book, they are certainly akin to the '*personal experience narratives around the intimate*' (Plummer 1995: 7, italics in original) which he analyses in that text. Plummer's specific investigations are of coming out, rape and individual recovery narratives, but the reproductive stories with which I am concerned are, in many respects, similar to those he examines.

Aside from these commonalities, Plummer's broader project in that book also provides helpful parameters for my investigation. Two key concepts deriving from this are highly relevant for understanding recent reproductive narratives. These constitute his attempt to generate a 'sociology of stories' and his notion of 'intimate citizenship' (Plummer 1995, 2003). He explains that he is less concerned with

> analysing the formal structure of stories or narratives ... and more interested in inspecting the social role of stories: the ways they are produced, the ways they are read, the ways they perform in the wider social order, how they change, and their role in the political process.
>
> (Plummer 1995: 19)

Plummer's larger project is one of aspiration rather than full realization in this book, but in struggling towards the larger project he highlights:

- the importance of community – story interactions;
- the need to understand community – story interactive *productivity* in its specificity;
- the role of sexual or intimate stories in the formation and mobilization of social/political groupings in the last decades of the twentieth century;
- the crucial role of the mass media in the generation and circulation of these narratives and the need for attention to specific media forms and their evolution;
- the importance of these interactions (stories–community–media) in the construction of subjectivities/identities.

100 *Telling stories*

These are the dimensions of his aspirations towards a 'sociology of stories' which seem most pertinent to recent reproductive narratives.

The conceptual hinge for Plummer's sociology of sexual stories is his notion of 'intimate citizenship'. Writing in the mid-1990s he presents this term by establishing the context for the recent emergence of new forms of politics:

> What has become both visible and practical over the past two decades (although the roots go further back) is the creation of these new communities of discourse and dialogue championing rival languages, stories and identities which harbour the rights and responsibilities of being sexual, pursuing pleasures, possessing bodies, claiming visibility and creating new kinds of relationships.
>
> (Plummer 1995: 150)

He goes on to identify and label this: 'A new set of claims around the body, the relationship and sexuality are in the making. This new field of life politics I will call "intimate citizenship"' (Plummer 1995: 151).[2] As my preceding account of recent reproductive narratives shows, these fit well into Plummer's 'new field of life politics'.

Lauren Berlant shares Plummer's concerns with the transmutation of citizenship and with the recent 'nationalist politics of intimacy' (Berlant 1997: 7), which she contends came into dominance in the Reagan era in the United States. In *The Queen of America Goes to Washington City: essays on sex and citizenship* she assembles selections from and reflections on her 'archive' (Berlant 1997: 11–12) of Reaganite US popular culture that embody these new forms of public political life. Like Plummer, Berlant pursues popular narratives of contemporary citizenship and offers observations about the subjects at their centre. She posits that 'citizen-victims' (Berlant 1997: 1) are the dominating subjects on the recent US political landscape, that testimonials of pain are the pre-eminent political narratives, and she deplores the concomitant 'cultural politics of pain' (Berlant 2000: 33; 1997). She situates this thus:

> I would like to connect it to something I call national sentimentality, that is, a liberal rhetoric of promise historically entitled in the United States, which avows that a nation can best be built across fields of social difference through channels of affective identification and empathy.
>
> (Berlant 2000: 34)

While Plummer sees both potential and danger in this new politics, Berlant is more unequivocally critical. Her research on recent testimonial forms is in some respects more generalized and in others more specific than Plummer's, and this contributes to their rather different political evaluations of 'intimate citizenship'. Berlant focuses exclusively on the United States. While Plummer sees his analysis of 'intimate citizenship' as pertaining most obviously to the United States, he ranges more widely, including British (and possibly other) exemplifications.

As already indicated, his study is oriented around three particular types of narratives (coming out, rape and recovery) and he tracks the communities with which they are associated. Assessment of the orientations of these communities and their visions for change leads him to distinguish amongst manifestations of the 'politics of intimacy': from the conservative vision of individual transformation of the twelve-step programme (originally developed for dealing with alcohol addiction) to the more radical aspirations towards political change which feminism and gay activism have sustained. *The Queen of America* (Berlant 1997), by contrast, is focused on the Reaganite revolution and its impact on US culture. Berlant highlights the icons of 'the fetus, the child and the immigrant' as the haunting figures of US politics from this period and offers detailed case studies of their construction/mobilization (Berlant 1997). As Berlant sees it, the obsession with these pre-citizenry figures has resulted in the dominance of sentimentality in public culture and 'a politics that abjures politics, made on behalf of a private life protected from the harsh realities of power' (Berlant 1997: 11). Berlant concentrates on the dangers in the forms and regimes of (apparent) truth and justice instituted by the 'politics of intimacy'. She deplores the takeover of the public sphere by such a politics in the United States, identifying the way it 'overorganize[s] the terms of public discussion about power, ethics, and the nation' (Berlant 1997: 8), precluding other political priorities.

Berlant's foregrounding of the iconic power of the child and the foetus goes some way to accounting for the purchase of reproductive narratives in the late twentieth and early twenty-first centuries. Moreover, in articles published since the publication of *The Queen of America Goes to Washington City* she has delineated the features and mechanisms of recent sentimental politics which pivots around public personal testimony. This includes her acknowledgement that traditions of women's virtuous suffering in the family and feminist discourses of rights related to such discourses have been part of the structuring conditions for this new political culture. As she sees it, these are factors which 'enabled the political consensus that situates narratives of trauma on the ethical high ground above interest politics' (Berlant 2000: 34).

A distinctive facet of Berlant's take on the testimonial narratives associated with the politics of intimacy is her exposition of their legalistic regimes of truth. In a later article, she highlights the appropriation of legal rhetoric and the production 'twistedly' of 'the law's multiple genres – evidence, argument, and judgment' in such testimonials (Berlant 2001: 42). Effectively, she traces the emulation and condensation of legal modes in such popular forms of address, showing how they become surrogate legal technologies that revolve around the presumption that pain is 'the only knowledge there is, more eloquent than and superior to the law' (Berlant 2000: 44). She ponders the mechanics of such testimony and the subject positions they offer: 'Mobilizing the putative universality of pain and suffering, the testimonials challenge you to be transformed by the knowledge of what you cannot feel directly' (Berlant 2001: 44). Critically questioning the self-evidence and obviousness of such knowledge (Berlant 2000: 35), she insists that '*psychic pain experienced by subordinated populations must be treated as*

102 *Telling stories*

ideology, not as prelapsarian knowledge or a condensed comprehensive social theory' (Berlant 2000: 42–3, italics in original). Reparation for such pain should not be taken as the guarantor of justice in Berlant's estimation.

Bringing together Plummer's and Berlant's research around 'intimate citizenship' creates a rich theoretical reservoir for the following analysis of reproductive narratives in the last two decades.[3] Nevertheless, the discordances and tensions between their projects also merit attention. As I hinted on pp. 100–2, their methods and their political prognoses are quite different. The methodological core of Berlant's approach is detailed text and image analysis, while Plummer pursues his sociology of stories, explaining that he is 'less concerned with analyzing the formal structures of stories or narratives (as literary theory might)' (Plummer 1995: 19). In relation to texts, Plummer's is a much more broad-brush approach: he does cite and comment on specific texts but these are not detailed readings. He concentrates on his typology and mapping. It is not the workings of texts so much as the activities and ideas of the communities which enable their production and of the audiences which consume them that interest him. Establishing that there is a productive cycle between communities and sexual stories which mass media forms have facilitated and realized, Plummer sketches the emergence of affiliated identities and social groups.

Plummer's and Berlant's projects are both inspired by feminist, gay and queer politics, but offer rather different prognoses concerning the flooding of the public domain by sexual stories. Plummer's tracking highlights specific types of testimonial narratives as positive – coming-out narratives and rape stories, associated with gay and feminist communities. His typology provides the basis for distinguishing between sexual stories that are oriented around political rather than individual change. Hence, he does not totally disavow the politics of intimacy. Berlant is more critical as she moves between textual readings (much more detailed than Plummer's), expositions of related regimes of truth and her perceptions of the degeneration of the US political climate which she argues has been generated in its wake. Their different methodologies are entwined with these different political assessments. Plummer is interested in audiences/listeners as active agents in the making of stories. His optimism about the potential of sexual stories derives from his positive perspective on feminist and gay communities of the late twentieth century. Berlant sees texts (particularly the right-wing ones she analyses) as powerful, because of the discourses they mobilize and the regimes of truth they instantiate. She does not consider the agency of audiences: the implication is that powerful and popular discourses render them powerless. In a sense, the insightfulness of her own readings mirrors the political power she attributes to the discourses she brings under critical scrutiny.

My brief discussion of Plummer's and Berlant's research concerning the politics of intimacy and the testimonial narratives through which it has been manifested provides a theoretical prism for my consideration of what has happened to 'I can't have a baby' stories since I first began to study these in the late 1980s. In the following sections of the chapter I examine the *resilience, proliferation* and *differentiation* of these narratives over that period.

Resilience

As my examples suggest, the 'I can't have a baby' stories have proved to be a remarkably resilient narrative genre. In the countries in which I have lived since the early 1980s (Britain, Canada and the United States), the media (including newspapers, magazines, television and film) have all circulated versions of this narrative. Narratives of women (or couples) who cannot have children, revealing their concomitant pain and suffering and speaking of their attempts to solve this problem, or third-person accounts of their plight have become classic tales of late twentieth- and early twenty-first-century Western life. I propose three reasons for the resilience of this narrative, which work out from the form, to the content and then the context of the stories.

These stories embody a classic and simple form of salvation narrative, as my examples illustrate. 'Amanda's story', in this sense, is a succinct and classic instance. In Ken Plummer's terms, they 'have their roots in classical stories of redemption and transformation', with ' a move from suffering, secrecy, and often a felt sense of victimisation towards a major change' (Plummer 1995: 50). The stark declarations concerning the problem of not being able to have a child and the exposition of the pain this brings engage the reader or listener. There then follows an account of the quest for a solution, generally involving encounters with scientific medicine. The reader/listener follows (often eagerly) this trajectory. The engagement with this quest sets up high expectations for the sought-for resolution of the birth of a healthy baby. The simplicity of this narrative line and the intensity of its potential for identification are key factors in the resilience of this narrative genre in the Western media since the mid-1980s.

A further feature of these stories, which also contributes to their resilience, is that, despite their generally highly personal nature, they are universalistic and involve no reflectivity about their specificity or context. As I indicated previously, they are spoken in the voice of everywoman, they presuppose and mobilize a seemingly universal desire to reproduce and they register this as an apparent right.[4] There is never any questioning of that desire or that right, nor any reflection about their location. In this respect, the declaration of pain and suffering (associated with the apparent inability to have a child) becomes an unquestionable truth, a form of 'prelapsarian knowledge' (Berlant 2000: 43) which legitimates the claim that resources (usually linked to modern science and medicine) should be mobilized to realize the *right* to reproduce. The highly personalized narrative form makes it unquestionable, whilst its universalistic framing elicits identification and sympathy. The intensely personalized form of these narratives and their power in evoking identification effectively fuses ' I' with everywoman, making these virtually unquestionable accounts.[5]

As Berlant (1997) has established, the foetus and the child have been key figures within the evolving forms of the politics of intimacy in the Western world. The 'I can't have a baby' stories hone the political resonance of these figures. These stories revolve around the expressed desire for a child of 'one's own'.

104 *Telling stories*

Marilyn Strathern has drawn attention to the significance of this trope from an anthropological perspective, referring to the 'colloquialism' as, 'an interesting fusion of whom one identifies with and what one owns against the world, which points to desire or preference' (Strathern 1992b: 26). As noted previously, she identifies 'the widely shared cultural assumption that persons desire children of their own' as 'Euro-American' (Strathern 1992c: 156).

Strathern's anthropological identification specifies the context and denaturalizes the truth claims circulated in these particular reproductive narratives. But it is also crucial that, as befits the recent politics of intimacy, these narratives render these *not* as expressions of desire but as *claims based on rights*. Hence, a further reason for the resilience of this narrative form is its rendering of this distinctive Euro-American cultural assumption into a plea for justice.

Finally, and rather more speculatively, I suggest that the resilience of this narrative form relates to its capacity for containment of broad anxieties about reproduction. These stories begin with a difficulty in reproducing and end with the resolution of this difficulty. It may not be too farfetched to suggest that they can function as reassuring moral tales about a range of forms of reproduction – individual, social and political. In this sense, these narratives are vehicles for airing, mediating and containing anxieties about reproduction in the contemporary West. They reassure listeners and viewers that the Western world has the technology (in both the literal and metaphorical sense) to continue to reproduce itself (literally and symbolically, individually and globally). In this respect their social and political significance is similar to that of the spectacles associated with the space race, associated with Winterson's (2001) comment quoted on p. 94. Moreover, the focus on these narratives may displace other less contained and more problematic stories about reproduction in the Western world, including those about child poverty (see Haraway 1997: 202–12).

Proliferation

It is not just the resilience of the 'I can't have a baby' narratives; it is their proliferation over the last few decades which has been so remarkable. More and more of these stories have been circulated and given a wide public airing. This could be taken as an instance of the 'discursive explosion' or 'incitement to discourse' around sexuality that Michel Foucault identified:

> Western man has been drawn for three centuries to the task of telling everything concerning his sex; ... There was installed ... an apparatus for producing an ever-greater quantity of discourse about sex, capable of functioning and taking effect in its very economy.
>
> (Foucault 1979: 23)

However, like Ken Plummer and others, I am suspicious of Foucault's sweeping assessment regarding such patterns. Plummer observes:

It is not the case that everything can now be said about sex [or indeed reproduction] ... Yet his [Foucault's] account neglects the rise of mass media in all its diverse forms, and it provides little space for the generation of *particular kinds* of stories *at particular moments*: it is all strangely undifferentiated.

(Plummer 1995: 122–3, my emphasis)

Like Plummer I maintain that *detail* and *specificity* are important here. This involves noting the growth in testimonial cultures, particularly in North America, but also in Britain, that have evoked the 'narratives of intimate life' of the late twentieth century which Plummer and Berlant have studied. As Plummer registers, the development of specific media forms appropriate for communicating such narratives is also important. In the case of the reproductive narratives I have been examining, these include (to mention just some of the most obvious forms): television and radio talk shows, 'docusoaps' and other forms of television which 'expose' personal life. These are extremely popular, distinctive media forms of the late twentieth and early twenty-first centuries which entertain through the promise of personal revelation. They have also provided platforms for the airing of apparent injustices related to 'the politics of intimacy' (Plummer 1995). They have proved to be media forms highly appropriate for the production and circulation of 'I can't have a baby' narratives. So, for example, the 'showdown' on the *Oprah Winfrey Show* (March 2001) which featured two couples contesting custody of twins whose births had involved an internet surrogacy arrangement is but one of countless examples of media generation, airing and proliferation of these stories.

The generic personal revelation media programmes have not been the only disseminators of the reproductive stories considered here. As 'Amanda's story' indicates, such narratives have sometimes provided the framing or structuring device or lead-ins for expositions about developments in reproductive technoscience for popular readerships and audiences. Indeed, it has become standard practice for television programmes about developments in biomedicine and technology to use these stories in this way. Such programmes often employ state-of-the-art visual technologies to render visible body parts and processes previously inaccessible to the naked eye and/or the lay viewer. They also celebrate the achievements of contemporary biotechnology. In these high-technology presentations, as in the more prosaic newspaper exposition considered previously ('Amanda's story', p. 97), personal narratives about reproductive difficulties provide the human-interest framing for such expositions. The internet has become another medium for the production and circulation of personal reproductive narratives. Sherry Turkle notes that the use of the internet has rendered the 'computer an intimate machine' (Turkle 1996: 26), a vehicle for constructions and reconstructions of self (Turkle 1996: 177–209). The internet has been an important site for the advertising of surrogate services and gamete exchange and for stories about reproductive difficulties. Chat lines have been crucial locations for the generation of reproductive stories. Talk shows and the internet are perhaps the prime contemporary media of the politics of intimacy and these have been important vehicles for the production and circulation of 'I can't have a baby' stories.⁶

The development of appropriate media has not been the only factor in the proliferation of the reproductive narratives which I have been tracing over the last decade. Plummer's commentary indicates that it is necessary to acknowledge and analyse specificity – the emergence of 'particular kinds of stories, at particular moments' (Plummer 1995: 123). As Margarete Sandelowski notes, '[i]nfertility became newsworthy in the 1980s' (Sandelowski 1993: 7). This brings us to another crucial condition for the proliferation of these narratives: the concerted excavation and generation of difficulties around human reproduction that have preoccupied Western science and medicine since the late 1970s. As I indicated in Chapter 5, the favoured term for reproductive technologies is now 'assisted reproduction' and more and more women and couples have been diagnosed as requiring technoscientific assistance. The elaboration of the ways in which reproduction can go wrong has made so-called 'natural reproduction' (that is, reproduction which does not require technoscientific intervention) exceptional. In Chapter 5 I noted Sarah Franklin's study of media representations of NRTs in which she observed that 'the need for scientific assistance to human reproduction' (Franklin 1993: 528) is now widely and insistently highlighted. Fears about infertility and media portrayals of conception as a remarkable achievement, as she explains, make it so that 'the necessity for technological assistance ... comes to be seen as product of nature itself' (Franklin 1993: 40). It is in these circumstances that 'I can't have a baby' stories have proliferated: more and more women (and couples) have formulated and broadcast their stories about difficulties having children. These stories beget more stories, media forms attuned to their resonance circulate these, and their dissemination confirms that reproduction without difficulties (particularly without technoscientific assistance) can no longer be presumed to be the norm.

Differentiation

Thus far I have been concerned with more-of-the-same patterns relating to the reproductive narratives that I have traced. Plummer refers to '*personal experience narratives around the intimate*' (Plummer 1995: 7, italics in original) as 'proliferating stories, multiplying stories, dispersing stories' (Plummer 1995: 78). My illustrative narratives hint that these stories can be embedded in different forms: diary documentation or stark factual reports in newspaper articles, but, such narratives have also been structured into plays, films, radio and television soap operas, and many other forms (see note 6). Moreover, disruptions of the established storyline (reproductive problem-identification of technoscientific solution–resolution in the birth of a child), including representations of unsuccessful medical/scientific procedures, have also been given limited airing (see, for example, Carpenter 2006). In addition, there may be some warrant in extrapolating from my examples to observe a shift in the technoscientific pivot of these stories in the last few years. In the 1980s and 1990s, IVF was the procedure which provided the transformational element in the stories of reproduction. As the narratives considered here suggest, PGD is increasingly the technological focus of such stories. However, as Karen Throsby has recently observed, IVF does remain 'the core technology upon which ... new and controversial reproductive and genetic technologies are based' (Throsby 2004: 191) and, as such, it continues to garner media attention (see Carpenter 2006).

In addition to these notable differences, there have been crucial patterns of differentiation emerging *around* and *through* these reproductive narratives since the early 1990s. I use the term 'differentiation' to highlight two interrelated processes occurring during this period. First, there has been a differentiation (in terms of the social identities) of the speakers of such narratives. Second, and more complexly, these stories themselves have become *technologies for differentiation* – fuelling controversy and inviting social/ethical adjudication. I consider each of these forms of differentiation in turn below.

As Chapter 5 indicated, when I first began analysing 'I can't have a baby' stories the homogeneity of the speakers was striking, although seldom noted. In the 1980s and early 1990s, generally it was white (apparently) heterosexual, premenopausal and generally relatively privileged women in the Western world who spoke publicly of the pain and problems involved in not being able to reproduce their 'own' children. While these remain the predominant group in the profile of NRTs (see Throsby 2004; C. Thompson 2005), in the last decade or so a more diversified range of speakers have followed them into the public arena to air their pain and claim their rights to reproduce. The most prominent among these have been lesbian or single heterosexual women,[7] gay men and postmenopausal women.

In and of itself this constitutes a fascinating and complex pattern that requires detailed empirical tracing and scrutiny. This diversification has taken rather different trajectories in different national settings. This can be illustrated with reference to the claims of non-heterosexual-identified women regarding reproduction in the UK. Since at least the 1980s such women have sought access to NRTs as a mode of access to procreation and motherhood. In Chapter 5, the so-called 'Virgin Birth' controversy which erupted in Britain in March–April 1991 was cited as focusing on the rights of such women to NRTs (see Estling 1991). I noted there that the *Daily Mail* (in particular) attempted to generate popular opposition to bar women who were not in heterosexual relations from IVF treatment. Amongst other features, this was a fascinating episode which exposed women apparently attempting to pass as heterosexual in order to secure access to NRTs.

While passing as heterosexual continues (with various degrees of success) amongst some women as a mode of securing access to NRTs, 'out' lesbian women, single heterosexual women and gay men have made more public claims to reproductive rights and access to reproductive technologies during the last decade and a half. Although this has sometimes garnered condemnation, there have also been crucial productive (in the Foucauldian sense) dimensions to this contestation. For example, Mette Bryld (1991) maintains that the contestation of rights to new reproductive technologies brought the first public acknowledgement of the figure of the lesbian in Denmark. As Bryld's argument illustrates, contestation over new reproductive technologies and 'I can't have a baby' narratives have been key elements in the evolution of identities and negotiations over citizenship in late twentieth-century Western states. In a related commentary in her overview of developments in the United States, Charis Thompson refers to '[t]he normalizing of single, gay, and lesbian parenting in ARTs', which she sees as having 'become part of the somewhat increased citizenship rights of these family models in recent years' (C. Thompson 2005: 7).

The pattern of differentiation around reproductive narratives is striking in this regard. Once the monopoly of the voices of white, relatively well-off, heterosexual, premenopausal Western women and couples was broken, the political dimensions of the 'I can't have a baby' story became more obvious. Whereas those traditionally assumed to be entitled to reproduce were *not* socially identified or identifiable, but spoke as 'everywoman' (or 'every heterosexual couple'), these new voices were publicly identified and labelled. When non-traditional agents began to add their voices to the chorus of those speaking of their pain in not being able to reproduce, they were clearly marked. Voiced by these marked citizens, 'I can't have a baby' stories came to be identified as *political* narratives.

This brings me to my second and related sense of the differentiation of these reproductive narratives. 'I can't have a baby' narratives have been extraordinarily effective vehicles for opening debates about reproductive rights and resources in the Western world. They have occasioned the claiming, clarifying, the developing and contesting of forms of intimate citizenship in North America and Europe. In their wake have come calls for legal reform and regulation (for example around the use of the sperm of deceased husbands), technoscientific development and experimentation (including the treatment of postmenopausal women to sustain pregnancy), as well as adjustments and innovations in commercial and social infrastructures. This includes clinics for infertility treatment and regulative structures and agencies – such as the HFEA, established in 1991 in the United Kingdom, and also the emergence of so-called 'reproductive tourism'.[8]

These stories foreground particular nubs of controversy about reproductive rights and resources. A very interesting mapping exercise could be undertaken which would reveal sometimes overlapping, sometimes distinctive, national patterns around these narratives. In Britain, for example, controversies in the late twentieth and early twenty-first centuries have revolved six main issues:

- heterosexual coupledom as a requisite condition for claiming reproductive rights;
- the right to use sperm or embryos without partners' permission;[9]
- the 'ownership' of children;
- menopause as a natural marker of the cessation of women's reproductive[10] capacities and rights;

- maintenance of generational and kinship demarcations;
- permission to produce so-called 'saviour siblings'.[11]

Around these foci of controversy, hegemonic norms around reproduction have been rendered explicit and contested. New identities and relationships have been generated in and through the negotiation over rights and resources around which these stories pivot. These include 'out' lesbian mothers, coupled gay male parents, postmenopausal mothers, women who are simultaneously both surrogate mothers and generational grandmothers of the same child.[12]

Conclusion

This chapter has explored the evolution of the genre of popular reproductive narratives identified in Chapter 5 and which I have labelled 'I can't have a baby' narratives. Linking these to new forms of intimate citizenship, which Ken Plummer and Lauren Berlant have identified, I have sketched the interplay between identity formation, communities and media forms that have been associated with the evolution and circulation of these stories. The resilience and proliferation of this genre of narratives have helped to make reproductive science 'the rocket science' (Winterson 2001) of the early twenty-first century – the focus of the technoscientific imaginary.[13] Meanwhile, a complex pattern of differentiation has spun out from these narratives, unleashing challenging questions about rights and resources. From the outset, such narratives have constituted an embodied politics – a 'biopolitics' (Foucault 1979) – and situating them in relation to Plummer's and Berlant's research makes this obvious. However, it was only when non-legitimated, 'marked' subjects (including, most particularly, non-heterosexually identified women, gay men and postmenopausal women) began to tell these stories that their political nature gained widespread social acknowledgement.

The gender politics of these stories is complex indeed. The vision of technoscience meeting the needs of and responding to the desires predominantly of women would seem a progressive contrast with the sagas of the space race. Moreover, as Bryld (2001) indicates and I have echoed, these narratives have opened a new public space for lesbian women and gay men. Indeed, their claims to reproductive rights embedded in these narratives in some ways normalize and legitimate these identities. Moreover, these are predominantly feminized narratives of technoscientific salvation. Nevertheless, as Leah Wild's IVF diary illustrates, the use of the autobiographical voice is no guarantee that women remain at the centre of new reproduction practices and accounts of them. In fact the heroes of these accounts are, for the most part, male scientists and doctors: in this particular branch of science a few male 'pioneers' (such as Ian Craft, Robert Edwards and Robert Winston in Britain) have become celebrated public heroes.[14]

The productive and challenging consequences of these narratives make it difficult to tar them with the brush of Berlant's condemning analysis. There are, nonetheless, disturbing features of the form itself which Plummer's more equivocal assessment and sociological focus does not pick up. For this reason, I have tried to unpack the specific emotional mechanisms at play – the voice of 'everywoman', the fusion of the personal and universal – and to highlight *aporia* including those regarding class and ethnicity that have characterized the genre. Countless other questions hover around this cultural form: about the seemingly inherent right to one's 'own' child, about the preoccupation with reproduction, about the distribution of resources, about the limitations as well as the achievements of twenty-first-century biomedical technoscience.

The circulation of such stories has been a very significant factor in the normalization of NRTs and 'assisted reproduction': these stories help to generate need for the new technology and to proliferate more stories. Developments in this field of technoscience, in turn, beget more stories of assisted reproduction. High-level reproductive and communication technologies are the mediators of this productive interplay. Finally, it could be noted that, just as the stories and spectacles associated with the space race generated as well as allayed anxieties about global power in the West, reproductive narratives generate anxieties about reproduction while containing them through promises of technoscientific salvation. Whatever Jeanette Winterson might think, and due in no small measure to these popular narratives, in the early twenty-first century making babies has, in many ways, become socially and politically equivalent to rocket science.

Part III
Witnessing spectacle

7 National and international spectacle
Gulf War I

> Visual culture used to be seen as a distraction from the serious business of text and history. It is now the locus of cultural and historical change.
>
> (Mirzoeff 1999: 31)

This chapter is about Gulf War of 1991 (henceforth referred to as Gulf War I)[1] as a crucial episode in the genealogy of military technoscience and as technoscientific spectacle. Gulf War I involved intense personal, national, and international relationships forged with and through technology. It was also an episode in which technology became the medium for the negotiation and renegotiation of the world order. Taking Gulf War I as a case-study highlights a key dimension of life in the modern and postmodern Western worlds: the military as an important site and, through the state, agent of technoscientific innovation and investment.[2] Thus, this chapter addresses the long-term and mutually constitutive link between technoscience and war. Writing in the wake of Gulf War I, Chris Hables Gray noted: 'The connection between war and technoscience has long been intimate; now it is integral' (C. H. Gray 1997: 7). The technological dimension of this conflict was similarly underscored by General Norman Schwarzkopf (the US Commander-in-Chief in this conflict), who designated it 'technology war' (quoted from *Business Week* in C. H. Gray 1997: 37). As these comments indicate, Gulf War I entailed substantial investment (literally and symbolically) in technology and major state involvement and international collaboration with a technological focus.

The second dimension to this chapter is the exploration of Gulf War I as it was experienced by most citizens of the Western world – as spectacle. The emphasis throughout this book is on the integration of technoscience into the fabric of everyday life in the Western world in the late twentieth and early twenty-first centuries. Generally, this integration is so ubiquitous as to be mundane. Nevertheless, some contemporary encounters with technoscience – and the first Gulf War was such an instance – are encounters with the spectacular. Indeed, Gulf War I constituted a 'theater of global display' (Harvey 1995: 86) in which technology both *facilitated* and *became* the display.[3]

Attention to Gulf War I as technological spectacle involves examining how it functioned during the conflict itself and, as I shall suggest, continued to operate after this specific military engagement ceased, as a form of entertainment. The main encounter of the vast majority of Western citizens with this war was through the television coverage of this conflict. In this sense, the technological spectacle of Gulf War I was domesticated through the medium of television (Kellner 1992; Hoskins 2004). This television coverage also marked a continuation of the general pattern of interdependence between military and entertainment (particularly film) technology which Paul Virilio (1989; Virilio and Lotringer 1997) has demonstrated. Moreover, it fits within long-term, established traditions of war-centred Western popular entertainment culture (including novels, comics, newsreels, television news, films and so on.).

The features of Gulf War I elaborated thus far – substantial investment (economic, but also symbolic) in technology, its spectacular nature, and intense personal, national and international relationships forged with and through technology –were not new or peculiar to this military engagement. These features characterize much modern Western warfare and indicate why military technology is an important focus for the cultural studies of technoscience. Nevertheless, examining the specific forms of this war's manifestation in the 1990s enables explorations of patterns of continuity and transformation in the genealogy of these key technoscientific episodes. So, for example, while communication technologies have always been an important part of modern warfare, in Gulf War I mass media and computer and information technologies were of paramount importance. The intensified use of sophisticated information technologies distinguished Gulf War I from its genealogical antecedents.

Moving beyond this initial assessment, the discussion which follows argues that the cultural significance of Gulf War I in the West coalesces around three themes in which technology is central:

- spectacular nationalism and the instantiation of a 'New World Order' community;
- corporeal disappearance and haunting;
- the demise of the soldier hero.

The analysis of each theme has two dimensions. In the first instance, these features are related to technological aspects of the war itself and pursuing them crystallizes its distinctiveness in the genealogy of modern warfare. The themes are fleshed out, in part, through analyses of representations of war embedded in popular entertainment media that appeared in the Western world in the aftermath of Gulf War I. While these media forms become in this chapter the media of my analysis and are linked to technological features of Gulf War I, they were designed primarily as vehicles for entertainment. These media were instrumental in the playing out (in many senses) of the cultural significance of Gulf War I. Hence, the second strand of my analysis is my reading of these entertainment media forms in relation to the ongoing production of meanings around Gulf War I in the Western world. Through engagement with television coverage, videos, films, video games and other technological products, such as the ones discussed in this chapter, Western consumers became active producers of the meanings of Gulf War I both during and after the period of military engagement.

Each of the thematic sections that follow opens with an introduction to a media form or product which leads into a brief description of a media image or set of images. Cultural studies has been extended and challenged through the explicit designation of the field or sub-field of visual culture in recent years (Mirzoeff 1999, 2002; Sturken and Cartwright 2001).[4] As I indicated in Chapter 2, important feminist cultural studies of technoscience research emerged in dialogue with art history and film studies in the 1980s and 1990s. Nevertheless, the conceptual and theoretical challenges of fully engaging with visual culture remain formidable. One interesting small illustration of this can be found in Nicholas Mirzoeff's (1999: 13–22) discussion of the limitations of approaches to visual culture which seek simply to transfer the modes of reading developed for the analysis of print texts to visual analyses. I offer my descriptions and interpretations, mindful of the challenges in this field that others have begun to chart.

The technology of spectacular nationalism and 'New World Order' communities

> The presence of cameras in the field of operations does more than exert a constraint on military actions. It changes the focus of hostilities from the enemy's fielded forces to the civilian opinion at home which sustains the will to fight.
> (Ignatieff 2000: 192)

> The gap between the global and the local in the contemporary world is most effectively crossed by the visual image.
> (Mirzoeff 1999: 255)

The Gulf War is a double video package which appeared in 1996, co-produced by the BBC (British Broadcasting Company) and FRONTLINE, the latter being a consortium of public US television stations. This WGBH Educational Foundation video is comprised of an assemblage of news footage clips, which appeared as part of the original television coverage of Gulf War I, and 'insider interviews'. As its packaging advertises, the video 'reconstructs the events leading up to the war, the war itself, and its aftermath'. As a cultural product and form this video connects with the tradition of British World War II newsreels as well as with the very popular Hollywood war films of the 1980s and 1990s (which also became widely circulated videos and subsequently DVDs) (Kellner 1992).

The excerpt which shows the start of Gulf War I begins with a rather hazy and seemingly abstract image – the view from one of the US bombers preparing to attack Iraq in the opening engagement of this war. The voiceover commentary is that of a pilot communicating with military headquarters in preparation for the attack. The footage of the build-up to the attack is followed by a brief explosion of sound and light on the screen.

For viewers without awareness of its context, this image would be difficult to decipher since there is little to be seen on the screen. However, most viewers would have at least two sources of knowledge which could help them to decipher this segment of the video. First, it is framed by the preceding narrative which indicates the build-up to Gulf War I. In addition, most Western news-aware television viewers underwent forms of visual training during the Gulf War I conflict which predisposed them to *see* the Iraqi terrain through the eyes of the technologically aided US missile pilot who is providing the commentary. This orientation or predisposition is an example of what Barbara Duden designates as 'commandeered vision' (Duden 1993b: 574–8) – viewers seeing what they are being encouraged to see.

There is then a cut to General Norman Schwarzkopf reflecting about how he marked the start of Gulf War I which identifies and confirms the preceding excerpt as part of a narrative pertaining to that war. In fact, Schwarzkopf was himself recalling his own original viewing of a live television transmission similar to the initial segment of the video. He recalls that he said a prayer and then, in what he describes as a 'totally chauvinistic, but totally appropriate' gesture, that he ordered the playing of *God Bless the USA*. There is then a cut and, as a series of images of US military technology flashes on the screen, *God Bless the USA* is played.

This juxtaposition of the spectacle of high-technology warfare and self-consciously (at least for Schwarzkopf) nationalist sentiment was, I maintain, a feature of Gulf War I. My interpretation develops in dialogue with Michael Billig's (1995) fascinating study of recent nationalism in the Western world, mentioned in Chapter 3. Billig contends that the dominant Western view of nationalism (at the end of the twentieth century) associates it with fringe groups and extremism. He contends that this preoccupation has made it extremely difficult for liberal, or even some critical Western, thinkers to address what he considers to be far more insidious and ubiquitous manifestations of 'banal nationalism' in mainstream Western culture. These involve the personal and social cultivation of nationalism through a variety of daily activities, rituals and icons – including what Billig (1995: esp. 2, 39–43) characterizes as the recent flag culture of the USA. He is disturbed by the suspension of critical awareness about the quotidian nature of late twentieth-century Western nationalism, in the midst of the increasing pathologization of forms of nationalism associated with so-called extremism, linked to designated 'fringe' groups and nations. Indeed, Billig contends that the hegemony of 'banal nationalism' in the West is so pervasive that it is scarcely possible for critical scholars to address or even notice it. Hence, he observes the disappearance into the royal 'we', without any acknowledgement of the invocation of nationalism, of otherwise critical commentators. Moreover, he maintains that, by identifying certain groups as pathologically nationalistic, rhetoric is produced which 'suggests that those nations that oppose "us" are more than parochial competitors: they can be transformed into enemies of international morality', and he indicates that they are also construed as enemies of universal rationality (Billig 1995: 92). Billig sees recent US global dominance and aggression, as well as cultural hegemony, as sustained and reproduced through the mechanisms of this 'banal nationalism'.

Billig's analysis is important for those interested in the mechanisms of Western hegemony, international conflict and some problematic dimensions of late twentieth- and early twenty-first-century Western popular culture. In critical engagement with his study, my proposal is that since the 1990s in Western countries – the USA and UK in particular – there have also been crucial episodes of what could be termed *spectacular nationalism*: television coverage of Gulf War I being such an episode. My suggestion implies that some forms of recent Western nationalism have been spectacular and theatrical. This is the case, even if, as Schwarzkopf's account illustrates, these episodes can themselves be tied back into banal activities such as the playing of *God Bless the USA* and domestic television viewing. Indeed, the Gulf War of 1991 could itself be regarded as a technology of spectacular nationalism, in that it was the vehicle for the creation and enactment – the performance – of nations and nationalism.

Highlighting the technoscientific dimension of some forms of nationalism in the West challenges the persistent binarism which Billig sees as endemic in recent Western complacency about nationalism. It destabilizes the association of nationalism with apparently 'backward-looking' social agents and agencies. Moreover, the periodic US-led coalition attacks on Iraq which occurred in brief outbursts during the 1990s (including, for example, one in mid-December 1998), despite changes in personnel, involved re-enactments, or, as Jeffrey Walsh (1995a: 6) calls each of these, a 'simulacrum' of Gulf War I – televised, high-technology spectacles framed in the United States and Britain by nationalistic commentary. The roles of some of the key players were taken by new actors in the 1998 conflict: Bill Clinton replaced George Bush; Tony Blair stepped in for John Major. Nevertheless, in television broadcasts the new actors framed and announced their 'mission' in rhetoric which was every bit as nationalistic (if less self-consciously so) as Schwarzkopf's.

My contention is that television coverage in the United States and Britain made Gulf War I itself a technology of spectacular nationalism. In effect, I am suggesting that television was the facilitating technology of late twentieth-century Western (especially US, but also British) nationalism, much as print technology had been the facilitating medium in the formation of nation-states in Europe. As Benedict Anderson (1983) has shown, print technologies (employed in newspaper and novel production) were crucial in the forging of the 'imagined communities' and identities of the nation-state. Douglas Kellner (1992) has shown in considerable detail how the orchestration of the mass media (particularly television) by the US government guaranteed that US viewers *saw* Gulf War I as a nationalist challenge and triumph. More recently, Andrew Hoskins (2004), with particular reference to Gulf War I, has explored the generation of specific television images which shape and inform social memories of war.

My designation of the Gulf War as an instance of 'spectacular nationalism' realized through the combined agencies of military technology and television relates to David Nye's study *American Technological Sublime* (1994). Nye addresses 'the American public's affection for spectacular technologies' (Nye 1994: xiii). His thesis is that '[f]rom the first canal systems through the moon landing, Americans have, for better or worse, derived a sense of unity from the common feeling of awe inspired by large-scale applications of technological prowess' (Nye 1994: book jacket). As Nye explains, there is a strong US tradition of expectation that society is welded together by technological spectacles which temporarily create a sense of community.

Nye's survey of key episodes and sites in US technological history does not include war or military technology, although he does consider the viewing of atomic bomb explosions (Nye 1994: 232–6). Nye explains that he excluded 'the experience of battle' because it 'merits a study of its own' (Nye 1994: xvi; 298, n.10). Despite Nye's omission, Gulf War I was witnessed as a sublime technological spectacle in this established US tradition. As Schwarkopf's reflections indicate, it was viewed both by those directly involved and by television viewers across the United States who followed CNN and other television coverage as an awesome technological achievement which bonded Americans, in denial of division, around a sentimental nationalism.[5] In the 1990s war and communication technologies continued, but transformed, the tradition of the American technological sublime which Nye has traced. The physical assemblage of large groups of people was no longer required and through the agencies of sophisticated telecommunication technology virtual communities were drawn together by technological spectacle to share experiences of sublimity and concomitant nationalism.

In his wide-ranging introduction to visual culture, Nicholas Mirzoeff explores in a generalized way the notion of the 'televisual sublime', which he identifies as 'an event that very few could hope to witness in reality that seems to take us of out of the everyday if only for a moment' (Mirzoeff 1999: 100). His contention that 'if newspapers formed a sense of national identity in the nineteenth century, as [B.] Anderson [1983] has suggested, television has become the imagined community of the late twentieth century' (Mirzoeff 1999: 99) resonates with my specific observation about the role of television coverage of Gulf War I in the USA. In addition, Mirzoeff notes that 'there are certain moments when television offers a collective experience of the fragmented postmodern world' (Mirzoeff 1999: 99). His proposal comes out of his engagement with the anthropologist Arjun Appadurai's explorations of 'a new role for the imagination in social life'. Appadurai explains that

> [t]o grasp this new role, we need to bring together: the old idea of images, especially mechanically produced images ...; the idea of the imagined community (in Anderson's sense) and the French idea of the imaginary, as a constructed landscape of collective aspirations. ... The image, the imagined

and the imaginary – these are all terms which direct us to something critical and new in global cultural processes.

(Appadurai 1990: 5)

While I have noted the use of the notion of the imaginary previously (particularly by feminist scholars who have been influenced by Appadurai's work) (Chapters 5 and 6), this concept is particularly pertinent to understanding the significance of television coverage of Gulf War I. For not only was television coverage of this war an important occasion of spectacular nationalism (particularly for US viewers in the tradition of the American technological sublime as traced by Nye), it was also crucial in the attempts to forge a 'New World Order' (to invoke the terminology of George Bush Senior). While the term 'New World Order' was vague and had been used in different ways on various previous occasions after World War II, it was invoked pragmatically by Bush in the run-up to Gulf War I (particularly in his 'Toward a New World Order' speech to the joint sessions of the US Congress, 11 September 1990). The televisual spectacle of the bombing of Baghdad as broadcast by the US-based CNN provided opportunities in which viewers did, in Appadurai's terms, 'annex the global into their own practices of the modern' (Appadurai 1990: 5). They were positioned to do this, as it were, through the eyes of US pilots and with the sanctioned US government-oriented framing CNN provided. In this sense, this television spectacle not only forged and reanimated the US national community, but it invited viewers in Britain and other US-allied countries into a new imagined global community (Appadurai 1990: 5) – George Bush Senior's 'New World Order' – in which the United States was the dominant global actor. Of course, as with all audiences, not all viewers were compliant and some resisted this invitation to communal, nationalist and global engagement.

I have argued that Gulf War I was an occasion of technological spectacle witnessed at a considerable distance by a mass television audience (Kellner 1995; Hoskins 2004). US audiences watched the performance of high-technology military weapons. Their witnessing of what were, in effect, Baghdad lightshows fitted well with the long tradition of the American technological sublime. The framing and regulation of this performance by the US military via CNN and other US television companies rendered Gulf War I into a technology of US nationalism and extended the tradition of the American technological sublime (which Nye has traced). The international reach of CNN coverage, the form of this military engagement (including night bombings of Iraq), the nature of the broadcast visual images themselves and their framing also came together to invite a wider audience into a global community identified with George Bush Senior's 'New World Order'. Subsequently, the docu-entertainment video *The Gulf War* itself became a technological packaging of the cultural legacy of that war. It offered US viewers the possibility of privatized encounters with technological sublimity, nationalist animation and a particular version of global community. The domesticated viewing of this high-technology spectacle in daily television and private video viewing constituted new versions of the banal nationalism Billig (1995) has so acutely identified and condemned.

Gulf War I and its aftermath: disappeared corporeality and the elimination of national trauma

> The centrality of the act of injuring in war may disappear – the centrality of the human body may be disowned. . . . Visible or invisible, omitted, included, altered in its inclusion, described or redescribed, injury is war's product and its cost.
> (Scarry 1985: 80)

> To what end are we seeking an escape from bodies? What are we mourning when we flee the catastrophe and exhilaration of embodiment?
> (Phelan 1997: 2)

Steven Spielberg's film *Saving Private Ryan* (released in 1998) has an opening sequence which has generated much discussion and controversy. The scene re-enacts the arrival of US forces on Omaha Beach (France) during World War II. The portrayal is graphic: sound, as well as the visual imagery, bombard the viewer with the physicality of this military encounter. Once the combat scene begins, there is a long, indeed lingering, scanning of the destroyed and maimed bodies of US soldiers who have landed under relentless fire. Cindy Weber observes that 'the opening sequence of *Saving Private Ryan*, assaults its spectators with the realism of war' (Weber 2006: 31). Many viewers have found this opening disturbing. Despite this, the success of the film (labelled 'the top grossing motion picture of 1998' on its DVD version) and the subsequently released video and later DVD indicates attraction as well as aversion.

The representation of war in post Gulf War I Western films is intriguing. Paul Virilio (1989; Virilio and Lotringer 1997) is the critical scholar who has most comprehensively and convincingly demonstrated the interrelationship between the development of film and of military technology. Moreover, Douglas Kellner (1992) has identified a more specific connection between film and Gulf War I. He argues that the narratives of popular Hollywood films of the 1980s and early 1990s (including *Rambo* and others) quite literally set the stage for Gulf War I. In effect, most US viewers absorbed the spectacle of Gulf War I in the terms and through the images circulated by these films. These two important arguments show links between war and popular film relevant to my study of Gulf War I.[6] Nevertheless, my fascination with the appeal of *Saving Private Ryan* relates to another kind of connection between mass media and the technological features of Gulf War I, which revolves around the forms of warfare and of its representations (and how audiences respond to these).

One feature of late twentieth-century war films, the most celebrated of which is *Saving Private Ryan*, is what might be labelled their hyper-corporeality. The much-discussed, long opening scene of *Saving Private Ryan* is corporeal in the extreme. In contrast, media coverage of Gulf War I in the USA and Britain was orchestrated to minimize the viewing of bodily destruction, suffering and death. This war was designed to be 'hygienic' in this respect and it was represented as disembodied. For the Western nations that participated in Gulf War I, this was a war in which direct viewing and contact with the bodies ('the body') of the enemy and imagery and traces of bodily destruction were minimized.

This was the result of many levels and forms of orchestration which can be traced. One impetus came in the wake of the controversies surrounding possible US casualties in Gulf War I and perceptions about how the US government had lost public support during the Vietnam War (Walsh 1995a). Mindful of the impact of US deaths in the Vietnam War, US military strategists sought to deploy technologies which would minimize their casualties. So successful were they in realizing this ambition that it was widely reported that young men involved in military service in Gulf War I were actually safer in the conflict region than they would have been at home in the USA, where they were at risk of death from automobile accidents and urban murders (C. H. Gray 1997: 37). Furthermore, the investment in military technology which would minimize allied casualties was also manifested in the de-corporeality which was a feature of military encounters during this war. As Michael Shapiro has noted, 'the targets of violence were rarely available to anyone's direct vision and were hardly ever available for direct contact' (Shapiro 1997: 93). Even the deployment of ground troops was organized to limit physical encounter with the enemy and to destroy quickly any evidence of bodily destruction. US soldiers moved in large armoured vehicles which rendered them immune to physical combat and Iraqi fighters' fire. The following episode, described by Shapiro, typifies the conduct of US ground fighting during Gulf War I:

> At the outset of Operation Desert Storm the American army broke through the Iraqi front line using earthmovers and ploughs shaped like giant teeth mounted on Abrams tanks. Thousands of Iraqi soldiers were buried – some still alive and firing their weapons – in more than 70 miles of trenches. In the first two days of ground fighting in Operation Desert Storm, brigades of the 1st. Mechanized Infantry Division, 'The Big Red One' (the sexual innuendo implicit in the graffiti speaks for itself) used this innovation to 'clean out' the trenches and bunkers being defended by more than 8000 Iraqi soldiers.
>
> (Shapiro 1997: 83)[7]

Such modes of ground combat created, as Robert Fox (in the *Daily Telegraph*) described it, a 'battlefield without bodies, without any trace, without any memorial of what had happened' (quoted in Shapiro 1997: 84).

The de-corporeality of ground combat complemented, but was rather less celebrated than, the air offensive conducted by allied forces. Here the US military used computer technology in the so-called 'smart bombs' which were supposedly self-orienting and designed to achieve 'surgical strikes' on their targets.[8] In a general analysis of 'the formal terms used for bombing' Elaine Scarry contends that the labels '"surgical strike" and "strategic bombing," already contain within them an anticipatory account of the resulting injuries as beyond the purposes of those doing the bombing' (Scarry 1985: 75).

Technology ensured that for allied Gulf soldiers and airmen service in this conflict was de-corporealized, and, in fact, such technology facilitated minimization of direct visual encounter with the bodies of the enemy. Despite this investment in precision technology, or perhaps because of it, there was no body count of Iraqi casualties during Gulf War I itself or in its aftermath. As this indicates, it was not just those immediately engaged in the conflict who participated in the disappearing of the body of the enemy during Gulf War I.

This de-corporealization was carried into and indeed reinforced by the media coverage of Gulf War I in the United States and Britain. The US and UK governments had learned negatively from the Vietnam War and positively from the Falklands War about what was required to maintain public support for military engagement. Government control was the watchword and media coverage of Gulf War I was tightly regulated in the United States and Britain. Images of the bodies of Iraqis who had been injured or killed by attack became the main taboo in this carefully orchestrated public iconography of war. There was only one substantial broaching of this taboo in Western media coverage and this was in the aftermath of the bombing of an air-raid shelter in which civilian women and children were shown to have been injured and killed in the US attack. The controversy in Britain about the Sunday newspaper the *Observer* (March 1991) printing a photograph of emaciated, suffering and even dying Iraqi soldiers in retreat on the road to Basra at the end of the war also showed the power of this visual restriction. The Western iconography of Gulf War I was de-corporealized and, with the few exceptions considered here, this feature of the war was experienced directly both by the US and UK forces fighting in the Gulf and by viewers at home.

This iconography was underscored by the mode of Western media coverage which furnished 'bomb's eye views', realizing what Anne Balsamo labelled a 'disembodied technological gaze' (Balsamo 1999: 127) which made it difficult for Western audiences to register the violence of this war. In addition, at least one commentator has contended that the war photography generated by Gulf War I was continuous with this general feature of its form and representation. Jeffrey Walsh highlighted the dominance of 'abstract and fractional' (Walsh 1995a: 16) images in the photographic exhibition held in London to mark Gulf War I. He regarded this as breaking with established traditions of war photography, associated with seeking 'closeness to traumatic events' (Walsh 1995a: 16). He noted instead that the photography of Gulf War I provided images realized through remote, wide-angled lenses.

Elaine Scarry has traced the general tendency to deny and obscure the 'centrality of the act of injuring in war' (Scarry 1985: 80) in Western culture. She outlines the routes by which this crucial corporeal dimension of war disappears. These may involve the omission of reference to this aspect in formal and casual accounts or various forms of redescription of war. A final set of paths involves some acknowledgement of 'actual injury occurring in the sentient tissue of the human body' during war, but, she explains, this awareness is *'held in a visible but marginal position'* (Scarry 1985: 80, emphasis in original) by the employment of four different metaphors which she suggests designate it as a byproduct or less significant. Scarry's analytical focus is the linguistic register, and the inadequacy of linguistic representation is her main concern.

My analysis (relating specifically to Gulf War I) complements her argument by considering how the *visual* markers of war injury can also be hidden or obscured. Moreover, I have shown that, in its specific features, Gulf War I clearly extended and intensified the general pattern of denial of the bodily injuries and destruction of war in Western culture which Scarry has traced. In terms of Scarry's long-range perspective, it is perhaps particularly significant that the US–UK protocol for this war dispensed with what she regarded as the *only* (if woefully limited) public register of corporeal injury and destruction – body counts (Scarry 1985: 70).[9]

The disembodied Gulf War I can also be seen as the culmination of longer-term trends in the development of military technology. Hugh Gusterson notes:

> In recent history technological advances have multiplied the volume of damage done to human bodies in war but, ironically, the importance of bodies as targets has become increasingly marginal to warriors who now score success mainly in terms of territory captured, enemy weapons destroyed, or industrial infrastructure disabled. ... In our contemporary postmodern era of nuclear and smart weapons the unprecedented ability of commanders to destroy entire bodies is matched by a partial preemptive disappearance of the body from representations of war.
> (Gusterson 1991: 45)

For Gusterson, presenting the history of recent military technological development unfolds 'the story of the objectification and disappearance of the subjective human body' (Gusterson 1991: 46). However, he associates Gulf War I with a specific intensification of this long-term trend:

> The representation of the Gulf War to the American public, both by the media and the US government, ... was remarkable for the way in which it treated bodies as objects for mechanical enhancement, weapons as surrogates for the bodies of warriors and, above all, for the extraordinary visual and thematic absence of dead, maimed, mutilated, strafed, charred, decapitated, pierced human bodies in a heavily televised war which surely claimed at least 100,000 casualties.
> (Gusterson 1991: 49)

The disappearance of the body which was such a feature of Gulf War I was not sustained in its aftermath. The most obvious lingering and controversial marking of bodies by this war in the West has been what Elaine Showalter designates as 'an unexplained illness that has been named Gulf War syndrome [GWS]' (Showalter 1997: 133). Controversially Showalter has contended that the sources of this malaise are psychic and that there is continuity between GWS, war neurosis, male hysteria and post-traumatic stress disorder which, she claims, together constituted something of a hysteric epidemic in the 1990s. Others have maintained that GWS is a physiological, somatic disorder. Whatever the explanation of its origins, GWS brought into visibility the suffering of Western bodies as a result of combat during Gulf War I. Scarry reflects on the physical records of war injuries when she observes that 'the record of war survives in the bodies, both alive and buried, of the people who were hurt' (Scarry 1985: 113). In this sense, GWS is part of the genealogy of what she calls the 'physical signs ... [that] suggest in ghostly outline how deep, how daily, how massive is a population's experiences of the residues of war' (Scarry 1985: 113).

Indeed, returning to the clip from *Saving Private Ryan* with which I began this section, it seems clear that the repressed corporeal trauma of Gulf War I came back to haunt the Western world in more than one way. The most obvious form of this haunting was and continues to be Gulf War Syndrome. But there also seemed to be a strong element of reaction to the Gulf War I projections of warfare as disembodied, as the vividness of the visualization of war carnage and suffering and the popularity of *Saving Private Ryan* suggests.

The demise of the soldier hero

Pure War no longer needs men, and that's why it's pure.

(Virilio and Lotringer 1997: 164)

Masculinity has slowly emancipated itself from the warrior ideal.

(Ignatieff 2000: 188)

In 1999 an advertisement appeared for a video game – 'Gulf War: Operation Desert Hammer' – which enticed viewers to participate in heroic virtual combat by mobilizing a sense of US nationalist outrage at the non-resolution of that war. The central image of the advert was of a US flag as toilet paper which conveyed a crude message about national abjection and embarrassment in international conflict. The message was that the Iraqis (Saddam Hussein in particular) had 'wiped their noses in it' and the proposal was that by purchasing and playing this game, unlike the US government, the advert viewer/game purchaser would not 'kiss ass'. The game offered the player opportunity to realize the nationalistic heroism that Gulf War I itself had not delivered.[10] The appeal of the advert and of the game revolved around the figure of the soldier hero.

The soldier hero has been a key figure in Western military history, Western masculinity and Western nationalism. In the introduction to his impressive study of this figure in British imperialist public and private imaginings, Graham Dawson observes:

> The soldier hero has proved to be one of the most durable and powerful forms of idealized masculinity within Western cultural traditions since the time of the Ancient Greeks. Military virtues ... have repeatedly been defined as the natural and inherent qualities of manhood, whose apogee is attainable only in battle. Celebrated as a hero in adventure stories telling of his dangerous and daring exploits, the soldier has become a quintessential figure of masculinity.
>
> (Dawson 1994: 1)

Dawson notes that such heroic narratives have been closely linked to the discourses of nationalism since the emergence of the nation-state. At the centre of such narratives (in both documentary and fictional form) is the masculine figure facing life-threatening dangers in combat on behalf of his nation.

I have already hinted at ways that Gulf War I undermined the hold of this iconic figure. The efforts made by US military strategists to minimize the risk to Western lives had this consequence. I shall now sketch some of the ways in which the specific technoscientific forms of Western engagement in Gulf War I and its media representations destabilized the hold of the soldier hero as a gendered icon of Western nationalism. There were four main factors involved in the dislocation of the soldier hero during this conflict: women's participation in the Gulf War, which precipitated what Carol Stabile refers to as a set of crises (Stabile 1994: 101–3) around war and its representation; the disappearance of the front line and direct combat; the focus on technology and technological achievement as the determining factor in the conflict; and the emergence of the figure of 'the cyborg soldier' (C. H. Gray 1989).

Women in Gulf War I

Gulf War I saw the largest Western military mobilization of women since D-Day, with 30,000 women engaged in this military conflict. This was the first time that women had served so visibly and on such a large scale in a war effort. As Nira Yuval-Davis contends: 'militaries and warfare have never been just a "male zone"' (Yuval-Davis 1997: 93). Nevertheless, female engagement in direct combat has been a powerful (if recently contested) taboo in the Western world. As Cynthia Enloe (1983), Carole Stabile (1994) and others have highlighted, holding this line constructs/enforces heterosexism and performs gender in the guise of the protector/protected binary. As Stabile notes, 'the distinctions between combat and non-combative positions are neither as rigid nor as predictable as the military would have civilians believe' (Stabile 1994: 115). This aside, the global circulation of images of women in military service during Gulf War I threatened this touchstone of Western heterosexist engendering. Stabile shows how US media commentators compensated by highlighting 'women's alleged inability to protect themselves' (Stabile 1994: 119) and attempted thereby to garner more support for the war. The coverage of female US military personnel who were taken as prisoners of war (POWs) became a site of acute interest and contestation given the unprecedented participation of women in this war. Tales of heroic rescue and of ostensibly indicative uncivilized infliction of injury by the Iraqis on female victims recycled the traditional Western vision of the soldier hero. Nevertheless, the figure of the Western female POW also stood as a constant reminder that the heterosexist binary of protector/protected of women/combatants had been shaken. Overall, the circulation of images of female soldiers during Gulf War I constituted something of a crisis in Western iconography focused on the soldier hero (Stabile 1994: 101–3) and disturbed 'the naturalization of the construction of men as warriors' (Yuval-Davis 1997: 94).

Disappearance of the front line and combat

Man-to-man combat and direct, physical confrontation with the enemy were not widespread in or typical of Gulf War I. As indicated previously, earth-movers and other mobilizing technologies were employed to minimize such confrontation. In addition, television was used by the US military to reconnoitre the enemy, as Michael Shapiro explains:

> As the Gulf War progressed, it became apparent that the identification of friends and foes relied on media coverage (in the broadest sense) not only for television viewers but also for combatants, for the targets of lethal violence were glimpsed primarily on video devices and were rarely available to direct vision.
>
> (Shapiro 1997: 79–80)

Shapiro contends that the dependence on such mediating technologies 'rendered the relationships between viewing and fighting subjects complex, for the targets of violence were rarely available to anyone's direct vision and were hardly ever available for direct contact' (Shapiro 1997: 94). In short, direct encounter with the enemy, which has traditionally been identified as the quintessential activity of the soldier hero, was pre-empted.

Heroic technology

Through following television coverage, Western viewers were dazzled by technological rather than personal military performance during Gulf War I. It is in the wake of this war that Virilio and Lotringer offer their assessment (quoted on p. 124) that such 'Pure War' (their term) no longer needs men (Virilio and Lotringer 1997: 164). C. H. Gray (1997) offers a somewhat different perspective maintaining that, from the late twentieth century, the soldier had been usurped as the hero of war, to be replaced by the technical specialist (including the strategist). Gray associates the displacement of the soldier hero with a new instability in the gendering of war and with a concomitant feminization of the work of soldiers. Despite this, he argues that the heterosexist gendering associated with war has been maintained through the masculinization of technology and condensed into military technological power/knowledge. As Gray sees it, heroic military power is still masculine, but it is increasingly *technologically* rather than *corporeally* so.

Cyborg soldiers

The Gulf War I soldier was not merely influenced by and displaced by weapons technology, but also technologically modified. C. H. Gray (1989) has documented the emergence of the 'cyborg soldier' who entered onto the world stage during Gulf War I. Drugs to 'make the pilot's body compatible with the night-visioning equipment in the fighter planes' and anxiety depressants were some of the ingredients in this transformation. Of course, in relation to drug modification, cyborgian soldiers were not a totally new phenomenon, as Pat Barker's (1996) trilogy of novels about World War I reminds readers.[11] Yet it was Gulf War I which intensified and extended these military technological regimes and which made the cyborg soldier a more prominent public figure.

As this suggests, the technoscientific nature of Gulf War I and its media representations destabilized and dislodged the figure of the soldier hero in the West. This war intensified and revealed currents of change in Western military organization and operation that had been in play before the outbreak of this conflict. Hence, my sketch of the elements in this dislodging can also be related to Nira Yuval-Davis' argument that the recent increased recruitment of women to the military in the West 'was aimed at transforming the military service from a citizenship duty into a "job"' (Yuval-Davis 1997: 98). This overview links the demise of the soldier hero to the other features of Gulf War I explored in this chapter. Traditionally, as a focus for nationalist sentiment the nub of the power of this figure resided in his risking his life for his country. Benedict Anderson (1983) contends that it was precisely this risk-taking which was pivotal in earlier mobilizations of nationalism in Europe. Of course, as I have argued, Gulf War I minimized such risk and, for the West at least, produced few such heroes.

The intense technological orientation of Gulf War I, the wide circulation of representations of women's participation in this conflict and the absence of images of death all rendered the soldier hero a shadowy presence during this war. Nonetheless, the figure of the soldier hero was not easily removed from the Western imagination. The advertisement for the Gulf War video game (and indeed the game itself) kept this traditional figure alive, inviting would-be purchasers to participate in simulated enactments of traditional soldier heroism and of crude nationalistic revenge and pride.

The Gulf War computer game is a high-technology reminder of the close affiliation of war with games in Western culture. Elaine Scarry comments: '"contest" is always attended by its near synonyms of "game" and "play," thus allowing war's conflation not only with peacetime activity but with that particular form of peacetime activity that is least consequential in content and outcome'

(Scarry 1985: 82). Scarry teases apart the fusing of *war* (as a particular form of contest) with *game* (as another, but distinct, subset of contests). She highlights the important differences between war and games, including that the latter demands only partial immersion and that they permit those involved 'to enter and exit from it freely' (Scarry 1985: 82). Scarry concludes that '[t]he severe discrepancy in the scale of consequences makes the comparison of war and gaming nearly obscene, the analogy either trivializing the one, or, conversely, attributing to the other a weight of motive and consequence it cannot bear' (Scarry 1985: 83). This clarification is an important critical intervention. Nevertheless, the widespread appeal of the cultural linking of war and games in Western culture is not easily dispelled. Moreover, Scarry's analysis revolves around formal and informal descriptions of war with their 'allusions either to specific peacetime games or to the generic attributes of the universal form of games' (1985: 85). She does not consider the double folding-in of this metaphor in manufactured war games themselves. Video and computer war games at once deny the dangers of the bodily injury and destruction which lie at the heart of warfare, whilst promising the simulated rewards of heroic soldiering. In these respects they have some affiliation with the war-time fiction that Scarry considers – albeit by offering a different corporeal and cerebral form of engagement. They continue the tradition of war-time adventure stories and films oriented mainly towards male readers and viewers who may be attracted to and identify with the soldier hero.

Conclusion

This chapter has explored the cultural significance of the technological dimensions of Gulf War I. The intense and continuous on-the-spot television coverage of this conflict in what was described (at the time and in its immediate aftermath) as 'the greatest media event in history' (Hudson and Stanier 1997: 209; Kellner 1995; Hoskins 2004) ensured the integration of this conflict into daily television viewing in the West (particularly in the USA).[12] The televised spectacle of Gulf War I became, I have contended, a technology of nationalism which was integrated into the domestic lives of many US and UK viewers and the mechanism for forging 'a New World Order'. My argument extends Michael Billig's challenges about the recent blindness to forms of Western nationalism by demonstrating that such sentiment can be mobilized not just around traditional activities (epitomized in flag display), but also around high-technology performance and spectacle. Indeed, in a related way, Tony Blair's pledges about equipping British schools for the information and communication technology (ICT) revolution or about making sure that Britain becomes a major international player in developments around genomic technology are, like the Gulf War I spectacle, occasions of nationalist mobilization. Like other manifestations of Western nationalism which Billig (1995) highlights, these episodes have generally not been subjected to the critical scrutiny of scholars concerned with the rise of nationalism.

As I indicate in this chapter and elsewhere in this book (Chapter 3), banal nationalism has taken many forms in the late twentieth and early twenty-first centuries in the West and this chapter shows that the Gulf War I spectacle was a key occasion for its cultivation. Nationalist sentimentality has not been confined to 'rogue' nations, fringe groups or even to the traditional ritualistic activities in the West which Michael Billig brings under scrutiny. The video clip in which Colonel Schwarzkopf recounted his viewing of the initial bombing of Baghdad that marked the opening of Gulf War I, and his acknowledgement that his orchestration of state-of-the-art military technology was marked by the playing of *God Bless the USA* as, in his own words, an 'act of blatant [and one might add, clichéd] chauvinism', is a reminder that banal nationalism may be generated through technological spectacles.

The second section of this chapter (pp. 120–4) explored the corporeal disappearance which was such a distinctive feature of Gulf War I and which was so carefully orchestrated through both coalition military and media technology. This was a manifestation of what Anne Balsamo describes as the 'very traditional cultural narrative' of the assumption of 'the possibility of transcendence' (Balsamo 1999: 128). Nevertheless, as I have shown, in the immediate and more long-term aftermath of Gulf War I this technological orchestration came undone, in part because of Gulf War Syndrome. This corporeal legacy has ensured that the impact of that war on physical bodies in the West has been registered and brought to public attention periodically in a highly contested way during the period since Gulf War I. Moreover, popular Western film production and viewing in the aftermath of Gulf War I seem also to have registered a strong reaction against the denial of the corporeal costs of war. The creation of and audience attraction to vivid portrayals of war's physical consequences – death and bodily destruction – in film and other popular media has been in direct tension with this feature of Gulf War I.

The final section of this chapter (pp. 124–8) considered the demise of the soldier hero in Gulf War I. Here I sketched the technological dimensions of this displacement, referring to new technical heroes, cyborgian soldiers, feminization and the deskilling of soldiers. The consideration of the displacement of the soldier hero in Gulf War I recapitulated the arguments of the preceding sections, because the risking of life in combat has traditionally defined this figure and because of the centrality of the soldier hero in European and US nations and nationalism.

The United States and Britain simultaneously mobilized military and media high-technology in Gulf War I. This chapter has drawn out the consequences of the interplay between these: nationalistic technological spectacle integrated into quotidian domestic television viewing, de-corporealized warfare and the demise of the soldier hero. The analysis spins out from brief analyses of indicative entertainment media – a segment of the PBS (Public Broadcasting System) documentary video *The Gulf War*, the opening sequence of *Saving Private Ryan* and the magazine advert for the video game *Gulf War*. These media testify to the established link between entertainment and military technology which Virilio

130 *Witnessing spectacle*

(1989), Baudrillard ([1991] 1995) and others have highlighted. Beyond this, these media forms were the vehicles for the continuous creation and recreation of meanings around Gulf War I in its aftermath in daily life in the West. In this respect, it is important to acknowledge the affinities between late twentieth- and early twenty-first-century watching of the documentary video of Gulf War I or the playing of the video game named after it and the long tradition of US and UK popular cultural forms oriented around war, including adventure fiction and newsreels. Nevertheless, as this chapter argues, it is also crucial to understand the specific significance of these popular entertainment forms and the interplay between technology and corporeality that characterized both the military enactment and the mediated performance of Gulf War I.

Coda

While the preceding analysis was originally undertaken during the 1990s in the immediate aftermath of Gulf War I, it was revised for this book in the early twenty-first century as the United States and Britain remain very much engaged in military conflict in the same region. The recent conflict is commonly referred to as *the* Iraq War, although it is also sometimes called Gulf War II. In effect, the United States, with British support, did return to what had been so crudely and scatologically portrayed as the USA's 'unfinished business' in the advert for the *Gulf War: Operation Desert Hammer* video game considered on p. 124. Indeed, some US and UK politicians, including President George W. Bush (Bush *Junior*) and some journalists echoed and embellished these sentiments in the run-up to and early stages of the more recent war. While there has been a rich repertoire of recent Hollywood and some international and independent film on wars since 1991 (Weber 2006), Gulf War I has garnered limited Hollywood attention. The comically satirical *Three Kings* (1999) did bring US operations in that war under some critical scrutiny, but it was not a great box-office success. More recently, Sam Mendes directed *Jarhead* (2005), based on the Gulf War I memoir of US Marine Anthony Swofford, a film which provides a sad, reflective postscript to US soldiering in that conflict.[13] None of the US casualties represented in this film are attributed to enemy engagement: masochistic training, masculinist bravado and friendly fire are shown as posing far greater threats. Indeed, the enemy really only appears in the most haunting scene of the film, which represents the US troops marching along the road to Basra (the 'Highway of Death'), where they confront charred Iraqi bodies and vehicles, marking the devastation of US bombing.

Likewise, Gulf War I has had little attention from critical scholars, who have turned their attention to the more recent conflict in Iraq, as well as to other recent theatres of war (particularly Afghanistan, Gaza, Lebanon, Darfur and so on). Nevertheless, the genealogical features of Gulf War I, including those I have sketched, remain important, not least because of the shadows they cast on the contemporary world. I offer here a few suggestions to substantiate my contention about this influence.

The early twenty-first century witnessed enactments of spectacular nationalism and Western global communalism through the medium of television during the coverage of the attack on the World Trade Center in September 2001. It could perhaps be argued that this subsequent spectacle was almost a complete reversal of the spectacle of bombings of Iraq which had engaged US and British viewers a decade earlier: in broad daylight (not in the dark); on US homeland, indeed iconic US territory (not a distant land); with the human destruction fully displayed (no surgical strikes or abstract lightshows); with television coverage inviting intense identification with victims caught in their mundane working lives (rather than with pilots attacking an invisible enemy). The US-led invasion of Iraq in 2003 was framed and justified by some coalition leaders in terms similar to those that sold the 1999 video game – as an initiative to 'finish off' the project begun in 1991, which would topple Saddam Hussein and his regime. Nicholas Mirzoeff (2004) opens his fascinating study of the visual culture of this recent conflict by exploring very different viewings of the television coverage of the more recent bombing of Baghdad in a sports club in Long Island, New York. Moreover, the figure of the soldier hero has returned to haunt the Western world, since the twenty-first-century terrorist who, in effect, makes his/her body into a weapon is both a cyborg and a soldier hero *in extremis*. In addition, the refusal to provide Iraqi body counts, begun during Gulf War I, has now been fully instantiated and institutionalized as a feature of US and UK military practice and government policy.

8 Techno-tourism in Florida
American dreams, technology and feminism[1]

A tale of two visits

In 1966 I boarded a Greyhound bus in Toronto for an almost three-day journey. I was bound for Florida to stay with my aunt, who was estranged from my Catholic family, in part, because she had married a divorced man. More recently, I have joked that this trip was my family's attempt to get rid of me. It does seem extraordinary from the vantage point of the intense awareness in Western societies in the early twenty-first century of the dangers facing (certain) children and young women that my parents would have sent me off, with little money, on a trip which involved stopovers in bus stations in some very rough parts of the United States, to arrive more than two days later, after midnight, at the other end of North America, to meet a relative no family member had seen for about five years! But I was a would-be itinerant worker, going to the gold-rush land and boomtowns around Cape Canaveral for summer work. My parents' endorsement of this adventure perhaps came because of their own background as migrant workers, from working-class families in a chronically economically depressed part of Canada: my mother had travelled from Cape Breton Island, Nova Scotia, to Boston to be a domestic servant; my father went to Europe in the armed services during World War II; then they both moved to Toronto, seeking employment. For them my trip was not extraordinary.

As it turned out, I got more than I bargained for in Florida, but I did not get a job. My parents' and my own hopes that I would make my fortune, or at least earn my keep for the next year of high school, were thwarted by the lack of public transport, crucial because my aunt and uncle (like so many others in that area) lived in an isolated trailer park. There were, however, other compensations. I walked onto what seemed to be the set of an American soap opera that, to use a 1960s phrase, 'blew my mind!' My aunt had constructed a new, much younger identity for herself, in which I quickly became complicit as I learned not to choke or protest as she referred to her younger sister (my mother) as her older sister. My new-found uncle worked at 'the Cape' (as it was called locally). A union man, he was handsome, with a macho veneer. Despite the bravado, I thought he was really a gentle guy – Jimmy Dean meets Ricky Nelson (an American pop singer of that era) – and he had at least one former wife. His sister was what I would now identify as an Imelda Marcos figure: she had a wall-size cupboard that was full of shoes. But more significantly for this chapter, she was secretary to the deputy director of the National Aeronautics and Space Administration's (NASA) operations at the Cape.[2] The scandal which was the buzz when I arrived revolved around her daughter, who was having an affair (this was the first time that I had knowingly met someone who was having one) with a foreman at the Cape. Then two trailers down the street there was Vicki-the-writer: she was my lifeline, providing me with a steady, if eclectic, stream of reading material. This included detective fiction, *The Complete Works of William Shakespeare*, as well as *The Art of Loving* (1962) by Erich Fromm. My aunt was convinced that Fromm's text was a dirty (that is, pornographic) book. There was also my uncle's nephew, who looked like a character (albeit a somewhat chubby one) from the popular 1960s American television programme *77 Sunset Strip*. He was well equipped for the part, with his convertible and his slicked back hair. He was the only single man I met that summer and, rather predictably, I fell in love with him. After all, he had been a member of the Green Berets[3] and he was my only reliable source of transport – taking me to mass (the drive-in cinema served as a Catholic church on Sunday mornings). He also worked at the Cape.

I spent much of that summer fantasizing about my hoped-for romance, deciphering the complexities of my aunt's new family and unpacking her fabrications about her old one, eating chocolate fudge ice cream, learning how to cook American-style fried chicken and devouring Shakespeare, Fromm and other less savoury texts provided by Vicki. The official high point of my visit came when my aunt's sister-in-law managed to get me a ticket to watch the Gemini II launch[4] from the official viewing stands. I wrote some very bad poetry after that, waxing unlyrical about all things American. It is perhaps not surprising that men and rockets figure prominently in that poetry: but there was no reference to my real American love – chocolate fudge ice cream.

Almost twenty-four years later, in spring 1990, I returned to visit my aunt and uncle during a year spent as a visiting professor at a small liberal arts college in Pennsylvania. By now my aunt and uncle were integrated into my family, my aunt having finally 'come out' as an older woman. Although this was again a family visit, I would probably have been classified as a tourist – coming, in part, to visit the by then well-established tourist sites of the EPCOT Center at Disney World, Orlando, Florida, and NASA's Spaceport, USA, at the John F. Kennedy Space Center. Ironically, although I passed as a tourist, this was also a work trip: I wanted to observe, reflect and write about these sites of technological tourism.

Although bits of my aunt's neighbourhood were vaguely familiar, this was a very different world from the one I had visited twenty-four years previously. Fortuitously and bizarrely, many of the same characters reappeared: the sister-in-law had retired from her NASA post and was living again with a rich, reactionary former husband. I did not check out her shoe wardrobe, but an estate in New York State and a big house and boat in Florida indicated that her consumption proclivities had continued apace. Her daughter, now installed in what I suspected was a truly boring marriage, made a brief appearance. Only Vicki-the-writer and the would-be-lover nephew were missing. The aura of boomtown and gold rush was gone: my aunt and uncle were ageing and marginal figures, without medical coverage, dependent on family benevolence and living in the shadow of more prosperous days. I saw them now as victims of American capitalism and managing in straitened circumstances, despite the stinginess of the US state and its wealthier citizens. In my returning incarnation, I was a much more suspicious, cynical character. Critical of heterosexuality, I was, nevertheless, still susceptible to its snares: I was recovering from recent episodes of non-romantic heterosexuality with lefty Brits – so-called 'new men' – whose escapades cast ex-members of the Green Berets in a rather positive light. The development of the Cape and the age of NASA and military boom had left their mark on the region but seemed to have waned. The shadow of the *Challenger* disaster (1986) was still palpable.[5] The future of NASA was shaky and the subject of considerable public debate: could the fragile US economy afford its extravagance? The focus of development in the region had shifted to another kind of gold-rush site – Orlando and, in particular, Disney World. This time I did not want to write poetry: I wanted to write critiques of all things American, most notably its science and technology, from a feminist perspective.

For my return visit there was no privileged access to official viewing stands to gaze amazed at American technological achievement and to marvel at the prospect of a technological adventure. I joined the masses and made my trek as something of a 'post-pilgrim' (Urry 1990) to two key shrines to technological achievement. In 1966 tours of the Kennedy Space Center were introduced and by 1990, when I returned to this part of Florida, Spaceport USA was a full-blown tourist site, consisting of a museum – the Gallery of Spaceflight – the Galaxy Center, with theatres, exhibitions and an IMAX cinema, a Rocket Cafeteria and Lunch Pad, and bus tours of the massive site. I also ventured to the region's new boomtown – Disney World – and I joined the crowds flocking to the EPCOT (Experimental Prototype Community of Tomorrow) Center. The full name was not often used, which was probably appropriate, since the acronym referred to the utopian residential community of Walt Disney's dreams and not to the EPCOT site I visited (Zukin 1991: 224–5).[6] When it opened in October 1982, EPCOT had two components – Future World (the Disney exposition of technological progress realized through corporate America)[7] and World Showcase (a Disney gesture to multiculturalism and global diversity).[8] Disney World, the most popular tourist site in the world, the paean to American middle-class, white culture and values, thus *incorporates* (literally, given the corporate sponsorship and perspective which sustains it)[9] an exposition of technoscientific progress. The two sites play off each other: on the one hand technology, progress and conformity in *Future Worlds* and on the other hand gestural representations of diversity and its packaged, tamed exoticization in the *World Showcase*.

Reflections on two sets of adventures in techno-wonderland, Florida

My narrative thus far has been constructed as a story of transformations, some of which were made explicit, while others remained implicit: from travel for work to tourism, class mobility, from the Cold War to the New World Order, the transformations wrought by capitalism and consumerism, involving transformations of identity, technological transformations and the transformations of feminism. I shall concentrate in the following analysis on the last two axes of transformation – technological transformations and the transformations of feminism.

It is crucial to note the particular setting of my story. This part of Florida has been since the 1960s a major site for the construction and dissemination of certain kinds of 'American dreams'. That is, it has been a site for the articulation of the dreams of certain kinds of Americans – predominantly white, middle-class, heterosexual Americans – and the realization of these, at least in representational forms, through technological development and material transformations. This area of Florida has experienced major technological transformations, beginning with the thousands of acres of swamp 'reclaimed' for Disney World. Technological triumphalism, faith in corporate America, investment in the nuclear family, endorsement of heterosexist hegemony and a racist insistence on the civilizing mission of white, middle-class America are the stuff that built this site. A NASA pamphlet which I brought back as a souvenir of my original visit declared: 'The area was inhabited by primitive Indians as early as 2000 B.C. ... American colonization began about 1821. ... During the construction of the Spaceport, Indian burial mounds and numerous relics of these eras were discovered and preserved' (NASA n.d. (a)). This imperialist mission was also apparent in the commentary of Dr Charles Fairbanks of the University of Florida quoted in a booklet on the Spaceport (from about 1966): 'This was one of the areas where Western civilization came to the New World, and now it is the area from which our civilization will go forth to other worlds' (NASA n.d. (b)).

I have to limit my exploration of this rich territory and I shall do this by focusing on me as a figure viewing technoscientific spectacle. After watching the Gemini space launch from the official viewing stands at Cape Canaveral, I wrote poetry which circulated clichés about this wonderful opportunity, about witnessing 'history in the making'. When I returned to that site in 1990, I participated with thousands of others in a much more dispersed collective witnessing of the triumphs of American technoscience (Nye 1994). There were at least four differences between these two opportunities for homage to technology which I can highlight. First, my 1990 visit to Spaceport USA inserted me into a carefully engineered 'mass' experience, in contrast to my rather privileged observer status (even if I secured my pass by happenstance) in the 1960s. Second, the dimensions of the technological witnessing during my second visit were much more elaborate and extended. Furthermore, my second encounter with celebrated technological achievement in this part of Florida itself required extensive technological intervention to create the spectacle or sense of awe. The

IMAX cinema and film and the 'person movers' rides/shows (such as 'Body Wars') at EPCOT were characteristic of these facilitating devices. In 1966 I simply stood on the stands and viewed the spectacle through my 'naked eyes'. Finally, I saw the display of technological triumph differently in my second visit because of my critical feminist perspective.

The first three differences suggest the features of late twentieth-century technological tourism, which, in the Disney World setting, involved the development and deployment of technologies for mass consumerism (including the IMAX cinema and 'person movers' rides/shows) which facilitated the celebration of technological achievement. Disney World's EPCOT and Spaceport USA literally cash in on 'the American public's affection for spectacular technologies', which, as I noted in Chapter 7, David Nye (1994: xiii) has brought under scholarly scrutiny. Indeed, both sites are oriented towards creating a sense of community around and through American technological accomplishment. Moreover, these sites are global tourist hubs which invite visitors from around the world to marvel at US technological achievement and to experience first hand the pleasures it can provide.

However, a more specific transformation which my narrative registered was of the way feminism altered my view of technological spectacles in Florida. This brings me to two reflective interludes, which I title 'Women watching I: viewing the spectacle of technoscience' (pp. 136–9) and 'Women watching II: feminist views of technoscience' (pp. 139–44).

Women watching I: viewing the spectacle of technoscience

Donna Haraway reminds us that '[w]omen lost their security clearances very early in the stories of leading-edge science' (Haraway 1997: 29). As noted previously, in her critical examination of the historical research of Steven Shapin and Simon Schaffer (1985) she highlights that, from the beginnings of modern science in the seventeenth century, extending into the late twentieth century, women were denied the status of legitimized 'witnesses' of science. This generalization has been fleshed out in a stream of feminist and feminist-influenced research since the 1970s, including Barbara Ehrenreich and Deirdre English's (1973a, 1973b) pamphlets on witchcraft and midwifery (see also Ehrenreich and English 1979), Caroline Merchant's (1980) and Brian Easlea's (1980a, 1980b) crucial investigations of the foundations of modern science, Margaret Rossiter's (1982, 1995) two-volume charting of women in science in the USA and David Noble's graphically titled study *A World without Women* (1992). Nevertheless, I want to linger on the image of women's *watching*, rather than *witnessing*, the spectacles which have constituted technoscience. I shall muse on the fact that I *was* given security clearance in 1966 and on the many women, men and children who since 1982 have eagerly bought passes (passports, tickets or otherwise obtained clearance to enter) into the EPCOT Center.

There has been a preoccupation with vision in much feminist analysis of science and technology – interrogating 'the male gaze' and the ways in which science (particularly the biomedical sciences) facilitates and effectively becomes the watching of women (Jordanova 1989; Haraway [1985] 1991a; Keller 1992; Cartwright 1995; see also Chapter 4 in this volume). Invoking the instances cited earlier of me and other women viewing the spectacles of technoscience, I would suggest that it is not just women witnessing or not being permitted to witness science or women being subjected to the scientific or medical gaze which merits critical attention. The phenomenon of women watching science also requires feminist exploration. Here I shall sketch a few striking scenes for consideration.

Donna Haraway (1997: 26–32) refers to Elizabeth Potter's reading of Robert Boyle's seventeenth-century text *New Experiments Physico-mechanical, touching the spring of the air* (1660), which describes experiments with air-pumps and provides an account of a demonstration, attended by upper-class women, which involved the evacuation of the air in a chamber containing birds that were killed by the process. Apparently these women interrupted the experiments, demanding that air be let in to rescue the birds. Boyle reported that, in order to facilitate such experiments, the men decided to assemble at night, thereby avoiding potential female disruption (Haraway 1997: 31).

Joseph Wright of Derby's 1767–8 painting *Experiment with a Bird in the Air-pump* shows women and children observing a male experimenter undertaking a procedure similar to that performed by Robert Boyle and which was analysed by Potter. This fascinating painting continues to attract viewers and interpreters in the early twentieth century.[10] What I find most intriguing about this painting is the evocative way it conveys both the interest and the unease of the female observers.

There is the related apocryphal story of the debate between Bishop Samuel Wilberforce and Thomas Henry Huxley on Darwinian theory held on 20 June 1860 at the Oxford meeting of the British Association for the Advancement of Science. This is widely regarded as a key moment in the history of science, not least because, as one commentator has noted, '[n]o encounter between science and religion has been more often described' (J. R. Moore 1979: 60). In the midst of this gentlemanly clash, a lady fainted and was carried off.

These are three distinct representations of women observing the spectacle of science. Two of these representations convey apocryphal narratives about women's excessive bodily reactions to the spectacle of science which resulted in men banishing them from or literally carrying them off the stage of science. In this sense, these representations illustrate female bodily reactions which could be and, it seems, were used to justify women's disqualification as potential 'modest witnesses' (Haraway 1997) of science. As these images suggest and as I have argued in Chapter 4, only *certain* women were permitted to see or be part of science in the making until well into the twentieth century. Indeed, tracing in which circumstances and locations and precisely which women do get a 'look in' in various historical periods and different national settings has been an important strand of research for some feminist historians of science. So, for example, class privilege was crucial in each of the instances considered above. Related to this, some important research has been undertaken on heterosexual coupledom and daughterhood as providing entrees into science for some women (see Abir-Am and Outram 1989).

In highlighting the figure of women *watching* science I intend more than a call for new research. Behind my cursory explorations is the claim that such moments *are* crucial in the making of science and in the making of gender. Hence, the figure of the woman viewing technoscientific spectacles must be juxtaposed with that other key figure – 'the modest witness' (Haraway 1997) of science.

As noted previously in this volume (especially in Chapter 4), this figure has been a powerful one in recent science studies and cultural studies of technoscience, invoked by Steven Shapin and Simon Schaffer in their much lauded book *Leviathan and the Air-pump: Hobbes, Boyle and the experimental life* (1985). Shapin and Schaffer contend that the foundations of modern science were established in the seventeenth century through Robert Boyle's pivotal role in generating three constitutive technologies:

> a *material technology* embedded in the construction and operation of the air-pump; a *literary technology* by means of which the phenomena produced by the pump were made known by those who were not direct witnesses; and a *social technology* that incorporated the conventions experimental philosophers should use in dealing with each other and considering knowledge claims.
> (Shapin and Schaffer 1985: 25)

Experimental philosophy spread through and with these material practices: these were the apparatuses for the production of what could count as scientific knowledge. As Donna Haraway explains, these 'three technologies, metonymically integrated into the air-pump itself, the neutral instrument, factored out human agency from the product' (Harawat 1997: 25). The gentleman natural philosopher and, subsequently, the scientist emerge as the 'modest witness' in these practices. With Boyle as the progenitor 'gentleman scientist' and his followers, including Newton (see Chapter 3), the white, European, middle- or upper-class man has predominated in modern science as 'the legitimate and authorized ventriloquist for the object world, adding nothing from his mere opinions, his biasing embodiment', with 'the remarkable power to establish the facts' (Haraway 1997: 24). His has been the 'culture of no culture', as Sharon Traweek (1988: 162) has memorably noted. Haraway observes: 'His narratives have a magical power – they lose all trace of their history as stories, as products of partisan projects, as contestable representations, or as constructed documents in their potent capacity to define the facts' (Haraway 1997: 24). The 'modest witness' thus effects a strict division between the technical and the political.

As I have noted previously in this book (especially in Chapter 4), the construction of the genealogy of the 'modest witness' has been a valuable contribution to science studies. Shapin and Schaffer (1985) drew attention to the politically and socially constructed nature of the Scientific Revolution. They highlighted its class components, even if they failed to address its gendered dimensions (Haraway 1997: 23–39). But the figures of 'excessive women' viewing the spectacle of technoscience illustrated in my narrative of my technological adventures in Florida and in the apocryphal representations of out-of-control women watching science in the making must be added to this picture. I am suspicious about the apparent *modesty* of scientific witnessing and would contend that too much investment in the figure of the modest witness may obscure the full and complex picture of how and where science is made with which I have been concerned in this book. Moreover, I am deeply sceptical about origin stories – even progressive origin stories (which Shapin and Schaffer's (1985) clearly is). As the narratives and images described earlier suggest, the modesty of scientific witnessing has always required another kind of watching and another kind of storytelling that was much less modest: popular observations of spectacles of technological prowess and tales of technological wonder. In fact, I would claim that the promise of technological progress lies hidden behind the documentation of the observations and experiments of the modest witness. Hence, the modesty of the scientific gentleman in his experimental mode as the 'ventriloquist of nature' can only be sustained by a more popular and non-cognitive relationship to technological promise. As Haraway observes, '[d]azzling promise has always been the underside of the deceptively sober pose of scientific rationality and modern progress within the culture of no culture' (Haraway 1997: 41).

My review of my own encounter with technological spectacle of the twentieth-century space race and of striking images of women viewing science in the making in seventeenth- and nineteenth-century Britain underscores the arguments of other science studies scholars (Cooter and Pumphrey 1994; Erickson 2005) who insist that making science is a much more dispersed activity than even its more progressive storytellers would have us believe. My specific contention here is that the practices and modes of those formally designated as the legitimate and legitimized makers of science are not always modest. Moreover, the more dispersed and diverse activities that could be characterized as 'watching science' are also part of its making.

These contentions reinforce the pertinence of the term technoscience which registers that the technological dimension of contemporary science is not simply an 'add-on' realized through commercial application, but rather that it is integral to its constitution.[11] Technological spectacle in specific forms and in its wide cultural currency has been crucial in sustaining twentieth-century science. The staged NASA launches, Spaceport USA and Disney World's EPCOT all inculcate expectations about technoscience and invite subscription to notions of technological progress.

Women watching II: feminist views of technoscience

I shall now switch my attention to another kind of viewing of technoscience – the overviewing which has yielded feminist perspectives on technoscience. While the preceding section of this chapter (pp. 136–9) proposes the foregrounding of the *female viewer*, the second part of my project examines the *feminist viewer* and feminist views of technoscience.

From the 1970s through the 1990s, feminism provided me and others with an alternative vantage point from which to view Western societies and Western science and technology in particular. The transformed 'me' who returned to Florida in 1990 saw the world through *feminist eyes*. I want to play out that metaphor and explore the experiential as well as the cognitive dimensions of this apparent transformation. When I returned to the Cape area of Florida in 1990 I was much more distanced and detached in many senses than I had been on my earlier visit. In fact, I toured both EPCOT and Spaceport USA, on my own, acutely aware of being outside the nuclear family and the heterosexual couple. Nevertheless, I was mindful that the real outsiders were those who did not get through the gates – those marginalized in and/or by the American dreams generated and mobilized at these spectacle sites. Indeed, like Alan Bryman (1995: 92–5, 128–9) and other critical commentators on Disney theme parks, I was struck by the overwhelming profile of the clientele, which consisted, then at least, of mainly white, middle-class American families.

But there was a further dimension to my distance: I had deliberately and self-consciously taken a *critical distance* from the technological triumphalism, the celebratory tone, the deference for the corporate-sponsored and state-sanctioned American visions of progress which were the preferred messages on offer there.

I shall label my vantage point – the platform from which I viewed EPCOT and Spaceport, USA – *feminist science critique*. Pursuing the theme of vision, I would say that I felt that my new viewing-stand facilitated a *clearer vision*. I was convinced then that I was able, through the tools provided by feminism, cultural and science studies, to see more accurately what science and technology really were. But there were other dimensions to my sense of clarity. It required detachment and distance – a kind of standing alone – what Elspeth Probyn (1996) might call being *Outside Belongings*, that was both pleasurable and, sometimes, a bit painful (lonely or isolating).

In fact, many commentators and researchers who write about Disney theme parks discuss their relationship to their attractions and spectacle in terms of whether or not they *distance* themselves from the 'mass' pleasures on offer at such sites. Michael Billig explicitly frames this in terms of ideology critique:

> Any writer who wishes to criticize the Disney phenomenon has a problem: what sort of tone, and indeed literary form, should be adopted? If writers enter into the fun of things, they risk losing the critical edge: the resulting voice will be that of the contented customer. On the other hand, if critical writers stand back from the masses in their millions, then they risk sounding superior. Their voices may resonate with disdain, as their prose looks down on uncultured, unthinking masses, wallowing like pigs in the troughs of superficial pleasures.
>
> (Billig 1994: 151)

Billig's reflections relate also to debates about the cultural critic's stance on popular culture which have had been a chronic concern of cultural studies since the early 1990s (McGuigan 1992). Theodor Adorno's (1991) dismissal of popular music has for some time been a reference point in these controversies and, as I suggested in Chapter 2, British cultural studies as it developed in the 1970s and 1980s took issue with the Frankfurt School's disdain for popular culture. However, Louis Althusser's (1970) account of ideology as pervasive in popular culture was influential in British cultural studies during this period. Althusser and his followers tended to assume that it took the powerful vision of insightful left intellectuals to 'see through' ideological configurations.

I have been sketching the way in which feminism had transformed my view of technoscience. For the most part, this was a view from *outside* technoscience and it depended on its outsider status. Conceptually, as Billig (1994) suggests, it was the mode of ideological critique which sustained this alternative vantage point. Such an approach raised a number of problems for feminists. First, like all forms of ideological critiques feminist critiques of science have an elitist dimension and constituted a vaguely avant-gardist politics, espousing some version of the claim that 'I can see what most people fail to see'. The political tensions about this elitism have been played out in second-wave feminist scholarship. Moreover, the prospect of distancing oneself from science generated particular tensions for some female scientists and, indeed, some feminist scientists, whose professional identities and allegiances were forged in and through science. Evelyn Fox Keller ([1987] 1999), Londa Schiebinger (1999) and other commentators have addressed this issue.[12] More generally, critical positioning may involve a highly rationalistic politics, linked to the assumption that engagements with science and technology are purely cognitive. This ignores the emotional registers of how technoscience works which I have foregrounded throughout this book: investments, hopes, identifications, pleasures and other forms of emotional ties. Finally, in many respects the mode of critique can operate negatively and it may sustain a rather pessimistic politics.

In global historical terms the ground was shifting beneath my feet when I made that trip in 1990: my intellectual and political viewing-stand was being dismantled. There are many world events that occurred in 1989–90 to which I can refer in making this claim: the destruction of the Berlin Wall, the transformations in Eastern Europe, the release of Nelson Mandela and so on. More locally (referring to my daily life, rather than geographical location), the mode of feminist critique of science was itself being reconsidered. Fresh perspectives were informing feminist thinking about science, including, most notably, Donna Haraway's *Cyborg Manifesto* ([1985] 1991a), and these were shaking the foundations of feminism's viewing-stand. Developments on the international political scene, poststructuralism and postmodernism, changes within feminism and other developments were blurring the picture, disrupting the clarity of vision. I became less sure about where I stood.

In the 1990s there were two notable feminist strategies for getting technoscience in focus, which I shall consider briefly. These were advocating a feminist 'successor science' and cyberfeminism. The feminist successor science project, as discussed by Sandra Harding (1991: 295–312) and others, involves attempts to reform the practices and principles of the established natural sciences in the wake of feminist critiques. Cyberfeminism, in its 1990s incarnations, revolved around the figure of the cyborg – the hybrid of machine and organism – as it emerged from cybernetic theory. It posited continuity between animals and machines which could be studied through the overarching framework of systems theory. In Britain, Sadie Plant (1997) was the best-known exponent of cyberfeminism. In Plant's version of cyberfeminism the symbiosis between a distinctively feminine mode of rationality and changes associated with information technology held out the promise of the defeat of patriarchy.

The attraction of these two feminist projects was, I suspect, that they provided new vantage points and positive strategies for feminists in their engagements with technoscience. In contrast to feminist science critiques, the advocates of these approaches acknowledged and, in some cases, celebrated the attractions of technoscience. In the case of feminist successor sciences, this was expressed through orientation around the pursuit of an intensified and reformed notion of objectivity and a reworked version of the Enlightenment project of progress through scientific knowledge (Harding 1991).[13] For cyberfeminism this has taken the form of enthusiasm for recent developments in information technology and their spin-offs.

There is much that could be said about these particular feminist technologies of re-visioning technoscience. Whilst acknowledging their appeal, I am also mindful of some of the limitations of these strategies. The prospect of a feminist 'successor science' seems a limited and limiting re-visioning in its accommodation with much of science-as-it-is. Moreover, given the entrenchment of establishment technoscience, the prospects for the realization of a feminist successor science – of a science which would be genuinely altered in response to feminist criticisms – seem, at best, remote.

Despite its widespread appeal, cyberfeminism has come under critical feminist scrutiny (Squires 1996; Adam 1997; Wajcman 2004: 66–77). Judith Squires contrasts the views of three key feminist analysts of technoscience – Shulamith Firestone, Donna Haraway and Sadie Plant – each of whom has been designated as 'cyberfeminist', noting that

> Firestone ends *The Dialectic of Sex* with a section called 'Revolutionary Demands'; Haraway writes her *Manifesto* 'to find political direction in the 1980s'; Plant, writing in Major's anti-political 1990s, has no sense of political project at all.
>
> (Squires 1996: 209)

Squires is unequivocal in her assessment of cyberfeminism, as espoused by Plant and her followers, describing it as

> the distorted fantasy of those so cynical of traditional political strategies, so bemused by the complexity of social materiality, and so bound up in the

rhetoric of space-flows of information technology, that they have forgotten both the exploitative and alienating potential of technology and retreated into the celebration of essential, though disembodied, woman.

(Squires 1996: 209–10)

Sarah Kember offers a similar criticism of Plant's work, describing it as 'technologically determined apocalypticism and biological essentialism' (Kember 2003: 178).

In my previous discussion of Firestone's views in Chapter 5 I identified her with an earlier phase of twentieth-century feminist strategies toward and studies of technoscience (McNeil 2001). As both Squires and Kember insist, Haraway's approach to technoscience is quite different from either Firestone's or Plant's 'cyberfeminism'. I share many of Squires' and Kember's reservations about the cyberfeminism espoused by Plant and her followers. It worries me because of its naturalization of gender differences and binarism, its ahistorical, rosy picture of both technology and women, and its denial of the need for political struggle or change. Not only does it uncritically circulate the 'myth of cyberspace that celebrates it as a gender- and race-neutral space of disembodied, democratic exchange' (Balsamo 1999: 15), but it assumes that technological change – particularly as realized in recent developments in information technology – will guarantee increased power for women. Holding out the promise of a feminist technological fix, despite considerable evidence of the traditional gender divisions instantiated within twentieth- (and twenty-first) century communications and information technology, such cyberfeminism is messianic in its vision.

Despite my dissatisfactions, these have been important movements within feminist analysis and politics of technoscience. These feminist viewing-stands which emerged in the 1990s were, I suspect, edifices constructed, in part, in reaction to the negativity associated with feminist ideological critiques of science and technology. Moreover, they have been fuelled by a democratic impulse to open science and technology to more women and to seize new openings and possibilities for women in the world of technoscience.

While some feminists have maintained a critical distance from science and others have seen benefits in becoming insiders in technoscience, some of the most effective recent feminist commentators in this field have employed striking visual metaphors to evoke the ambiguity of feminist positionings and to conceptualize fruitful strategies with regard to technoscience. Donna Haraway's ([1985] 1991a) cyborg is the most popular and widely adopted/adapted figure of this kind. The promise of this figure is well captured by Anne Balsamo:

> Cyborgs are hybrid entities that are neither wholly technological nor completely organic, which means that the cyborg has the potential not only to disrupt persistent dualisms that set the natural body in opposition to the technologically recrafted body, but also to refashion our thinking about the theoretical construction of the body as both a material entity and a discursive process.
>
> (Balsamo 1999: 11)

In a related exposition, the feminist anthropologist Emily Martin employs three images, 'the citadel, the rhizome, and the string figure', to 'allow us to picture the discontinuous ways science both permeates and is permeated by cultural life'[14] (Martin 1998: 24). Haraway herself uses the optical metaphor of 'diffraction' (Haraway 1997: esp. 33–4) to designate the yearning for somewhere else, for change, for transformation, whilst living with and, as she describes it, often appreciating and admiring technoscience.[15] I would construe a feminist strategy which registers a yearning for change and transformation whilst acknowledging our quotidian implication in the making of technoscience, as it were, as being witnesses of/to technoscience.

While I welcome these powerful and evocative analyses and images, my concern is about the seduction of the visual within feminist science studies: my fear is that we have been much better at playing with the figures than we have been at change in other realms. Perhaps a similar dissatisfaction lies behind Evelyn Fox Keller's reflection that 'feminist theory has helped us to re-vision science as discourse, but not as an agent of change' (Keller 1992: 76). Significantly, even in this critical evaluation the imagery of vision is employed. My reservations can also be illustrated by invoking visual imagery to pose the question: has preoccupation with textual and figurative re-visioning allowed us to glaze over the political working-through required to transform technoscience? Moreover, there are other questions to be asked, such as: who shares and can participate in this re-visioning? The play with visual imagery comes up against differences in tastes and cultural capital, suggesting that there are other stakes involved (for example the political strategies of avant-gardism versus other political approaches). Feminists have conjured and examined 'the male gaze of science', modest and immodest witnessing of science, both amazed and critical gazing at technological spectacles, cyborgs, the citadels of technoscience studies, and yearned for diffractions, but it has been far harder to politically transform technoscience.

Revisiting and re-visioning techno-wonderland, Florida

To return to my title – technology, techno-tourism, American dreams and feminism – this chapter has been about my troubling over and of:

- the relationship between the production (techno-triumphalism) and consumption (techno-tourism) of technoscience and the gendering of this divide;
- the attractions and hollowness of 'the American dream' (linked to the technological sublime, as Nye (1994) has demonstrated) and its fraught relationship to other more basic American dreams (for example of universal healthcare provision);
- the relationship between feminist figurations and feminist transformations in feminist technoscience studies.

In 1994 I returned to techno-wonderland in Florida. This was during another period spent as a visiting academic – this time in Canada. I had come to Nova

Scotia in an identity made possible by feminism, facilitated by the Canadian state education system and financed by the vestiges of this social-democratic, liberal state, as a visiting professor of women's studies from England and, hence, as a very privileged migrant worker. The ironies in this particular transformation were not lost on some members of my Canadian family. Once again, I took a trip to Florida in part to visit my aunt and uncle. This time, their ageing was noticeable and my uncle had a mouthful of rotten teeth: he had not had any dental coverage or treatment for many years. There was good news on this front, at least, as treatment was imminent, because his rich sister had agreed to pay for it. I was combining my trip with a holiday with friends. They were coming to Florida for a conference and for the first time in my life I was a tourist of academia as I sat in on this event. It was a management studies conference and it disillusioned me about the use and abuse of one piece of technology – overhead projectors. The mode of our subsequent holiday was much more postmodernist than any of my preceding visits: we did 'the nature trip' to the Everglades and we didn't quite cruise in that most postmodernist of locales, Miami Beach. When I got to the EPCOT Center I was told that much of the site is to be transformed, with new and more exciting attractions planned.[16] When I got to Spaceport, USA, I was mindful of new US–Russian technological collaborations: the space race has been transformed into extraterrestrial collaboration. The Toronto *Globe and Mail* had described this new phase of space exploration a few weeks before my visit in the following way: 'The U.S. shuttle Discovery blasted off at dawn yesterday with a Russian Cosmonaut aboard, opening a new space age free of Cold War rivalry. It's ... the first time that astronauts and cosmonauts have been launched in the same spaceship' (*Globe and Mail* (Toronto), 4 February 1994: 1). How am I to see this new technoscientific panorama? How might feminists transform it?

Postscript to my travelogue and pictures of feminist visions of technoscience

In this chapter I have used stories about my own encounters with technoscientific spectacle in reflecting about some of the visions and versions of feminist cultural studies of technoscience. Reviewing my own pictures, I am mindful that the field of feminist technoscience studies has proliferated and grown over the last three decades. Moreover, technoscience itself is continually changing. The two threads I wish to return to are distancing (including the insider/outsider dilemma) and critique. At the end of the twentieth century and in the first decade of the twenty-first century feminist science studies has taken new forms. I have highlighted some of the challenges for feminist cultural studies of technoscience in the early twenty-first century in this chapter, including the investment in ideological critique and the preoccupation with figures. As much of this scholarship and the studies presented in this book suggest, science and technology are not monolithic and, as Ann Balsamo warns, '[i]deological critique may be totalizing' (Balsamo 1999: 123). Stepping back from such critique, recently a number of feminist science studies researchers have deliberately eschewed intellectual/political positioning as 'outsiders', seeking fuller engagement with technoscience and technomedicine. This has been the strategy of both Annemarie Mol (2002) and Myra Hird (2004) in relation to biomedicine and biology, respectively. Sarah Kember takes her cue from Isabel Stengers' (1997) advocacy of 'risk', as she engineers her own 'close encounter between feminism [in fact a very particular cyberfeminism, influenced by Haraway's work] and artificial life' (Kember 2003: 176). Much of this research has been challenging and it has reinvigorated the field, providing new insights and fresh vantage points. Nevertheless, as I see it, feminist science studies analysis *does* require some *distinctive perspectives* from which to evaluate and assess the technoscience we scrutinize. As my opening narratives signal, technoscience permeates our quotidian lives and our life-stories. Hence, the critical capacity of feminist activism and scholarship has been crucial in identifying specific ways that modern technoscience has thus far *not* been adequate to feminist requirements and in suggesting how it might meet these more fully. It has been and no doubt will continue to be difficult for feminist science studies to deal with this tension between critique and enmeshment. But, as Margaret Atwood's character Offred in the *Handmaid's Tale* warns, '[p]erspective is necessary' (Atwood [1985] 1987: 153).

9 Conclusion

This is a book about science, technology and technoscience. It is about science studies, cultural studies and feminist studies and their coming together. It is about doing feminist cultural studies of technoscience. It is about the casting of particular heroes of science: Isaac Newton, James Watson, Rosalind Franklin, Barbara McClintock and Anna Brito. It is about the procreation stories that have been told and circulated at the end of the twentieth and the beginning of the twenty-first centuries, in the era of the new reproductive technologies. It is about the technological spectacle of the Gulf War of 1991. It is about technological tourism in Disney World's EPCOT Center and NASA's Spaceport.

In this conclusion I offer some final comments about the *doing* of feminist cultural studies of science and technology as explored and enacted in this volume. As I have emphasized throughout this book, cultural studies of science and technology comes in many forms and draws on a variety of resources. So there is no single origin story that explains the emergence of this field. The genealogical review provided in Chapter 2 was, as all such exercises necessarily are, partial and specific: focused on disciplinary routes into cultural studies of science and technology taken by some Anglo-American feminist scholars. I was surprised by the variety of disciplinary trails uncovered in this limited exposition and by the rich, complex tapestry of the field that unfolded. This exploration also confirms the importance of feminist contributions to the recent explosion of cultural studies of technoscience.

My review also demonstrates that, while there is nothing natural about disciplinary boundaries, the creative use, adaptation and transformation of the resources of academic disciplines has been crucial in the forging of cultural studies of science. Interdisciplinarity and trans-disciplinarity may have become 'second nature' to many cultural studies, feminist studies and science studies researchers in the early twenty-first century, but this has been realized as an achievement, rather than through innate orientation. Comparing and referencing other cultural forms (including fiction, art, popular film); interrogating the language, narratives and visual imagery of science; and borrowing and adapting concepts and theories (for example from anthropological kinship studies, literary studies, psychoanalysis, art history or film studies): these are the sorts of adventurous

148 *Conclusion*

undertakings which constitute cultural studies of technoscience and which enrich understandings of our technoscientific world.

In my conclusion to Chapter 2 (see pp. 22–4) I indicated the range of forms of analyses that can be assembled under the collective heading of *cultural studies of science and technology*. Given this pluralism, even eclecticism, it seems appropriate to reflect on some features of the analyses presented in the preceding chapters. These have included the employment of the framing notion of case studies which underscores the constructed and partial nature of any analysis. The use of this framing device is a practice which is also employed by science studies researchers associated with Actor Network Theory (Law 2004; Erickson 2005: 35) and in other areas of cultural studies (Berlant 1997; Pearce 2004).

The explicit foregrounding and employment of a diverse range of concepts and theoretical framings, including Benedict Anderson's (1983) 'imaginative communities' and Ken Plummer's (1995) 'intimate citizenship', is another feature of the studies presented in this volume. *Theory work* is an integral part of cultural studies of science and technology since these resources demand critical exposition and appraisal, translation and adaptation to be useful in specific investigations. Moreover, the foregrounding of this work challenges positivist or naturalizing modes in science studies analysis. Hence, one thread running through Chapter 3 was the exposition of how theory work and attention to the various makings and remakings of Newton have reoriented social and historical studies of this scientific hero, disrupting the obsessive search for the 'real' Newton.

The rhetorical experimentation with narrative form and the use of autobiographical writing (in Chapters 5, 6 and 8) are distinctive aspects of the preceding analyses. This is not to claim that such practices are totally exceptional, since there has been considerable experimentation with narrative and textual form in science studies writing since the late 1970s.[1] But the use of dream narratives and of the personal voice are not common in science and technology studies. Such modes were invoked in the preceding studies to register the ubiquitous, quotidian and pervasive engagement with technoscience in the contemporary Western world. They also signal that relationships to technoscience are by no means exclusively cognitive. These modes of representation derive from suspicions about 'the God-trick' (Haraway 1991b) and the voice from nowhere (Traweek 1988) which have been endemic within modern science.

This representational strategy is affiliated with second-wave feminist practices of 'giving voice' to women and feminism (Maynard 1994). But 'giving voice' is itself a tricky business. Thus, my forays were deliberately rhetorical and designed to highlight the complications feminists have confronted in eschewing essentialism and authenticity (Spellman 1988; Riley 1988; Probyn 1993), in articulating unease with testimonial forms (Berlant 1997, 2000, 2001), in confronting what Mary Evans characterizes as the 'impossibility' of autobiography (Evans 1998) and, more theoretically, in coming to grips with poststructuralism.

I signalled in Chapter 2 that questions about the positioning and representation of the analyst have hovered around the field of cultural studies of technoscience and they have lingered in a much hazier way over the broader field of social and technology studies. Accounts of ethnographic research in science and technology studies generally address methodological and ethical issues arising from the interaction between the researcher and the 'subjects' of the research (Franklin 1997; Rapp 1999; C. Thompson 2005). However, these and other acknowledgements of research practices and participants do not tackle the many dimensions of analytical positioning. Steven Shapin (1992) offers some interesting sociological perspectives on the changing conventions regarding analytical and political stands amongst practitioners in the discipline of social and historical studies of science in the twentieth century. More recently (as I noted in Chapter 8), epistemological orientations and methodological practices which entail the 'othering' of science, technology and medicine have come under critical scrutiny. The explication of the problems and dangers of such 'othering' has yielded original science studies research on biology and biomedicine (Mol 2002; Hird 2004). My unpacking of the attachment to ideological critique (Chapter 8), which has been an epistemological and methodological approach favoured in much feminist science and technology studies, addresses strategies of 'othering' and distancing in a related, but somewhat different, way. Nevertheless, while the dangers of 'othering' are palpable, it is important that science studies researchers also address the risks of uncritical embedding. The illusions of value neutrality and new forms of the 'God-trick' are easily reinstated.

Despite the ambiguity and variety of reference implied by the label 'feminist', it does eschew neutrality and signal address to the power relations in which science and technology are ensconced. For these reasons, the use of 'the f word' to identify practices within cultural studies of science and technology and within the broader field of science studies remains crucial for me and some other researchers. Feminism is the shadowy but significant force which I have sought to foreground in this book. I deliberately set out to investigate feminists' contributions to the burgeoning of cultural studies of technoscience and to document this work, mindful that, as Donna Haraway (1989, 1991a, 1992, 1994, 1997) and others (Delamont 1987; Wajcman 2000, 2004) insist, feminist contributions have not had the attention they deserve within science and technology studies.

I did not expect that feminism would emerge so vividly as a spectral presence in my case studies. Nevertheless, in tracing the making of scientific heroes (Watson, Franklin, McClintock, Brito) in popular biographical writing of the late twentieth century I detected feminism as the absent presence which haunts these texts. Likewise, I have noted that the obscuring of feminism has been striking in the midst of the proliferation of 'I can't have a baby' narratives and the excitement about NRTs in the late twentieth century.

Feminist activism, theory, research and scholarship have sustained my own analytical vision. Investigations of corporeality and embodiment (together with complementary studies of the obscuring and denial of the body), of heteronormative conventions and assumptions, and of the rendering of women either invisible or as subjects of surveillance are tropes of recent feminist studies that I draw on in the preceding chapters. These channels of research have made it possible to study both the promise of technoscience and how and why it often does not realize its promise.

Modern science has been intensely embroiled in the making of gender (and sexual) difference and in gendered and other forms of oppression. The foregoing studies highlight two of the many ways in which the male gaze has operated in and through technoscience: in the scrutiny of female appearance, even amongst high-achieving scientists of the twentieth century, and in the new forms of surveillance and self-surveillance associated with NRTs. Moreover, some apocalyptic moments in the restrictions on women as would-be witnesses of science have also been considered in this volume. Hence, it is not surprising that many feminist activists and scholars, in Anne Balsamo's terms, have been 'keeping watch' (Balsamo 1999: 97) on modern technoscience.

In my own watching I have been mindful that technoscience is pervasive in contemporary Western societies, but that it is not monolithic. Feminists have generated a marvellous repertoire of visual images of cyborgs, monsters, tricksters and cat's cradles to evoke women's technoscientific entanglements and agency (Haraway 1991b, 1992, 1994; Lykke and Braidotti 1996; Martin 1998) (although, as I hinted in Chapter 8, I fear that there has perhaps been an excessive investment in this imagery). As part of the movement of second-wave feminist activists and scholars who helped to forge feminist critiques of science and technology, I have become aware of the limitations of ideological critique, as I explained in Chapter 8 and noted on p. 141. Its rationalistic mode, its negativity and its reliance on distancing and 'othering' have been problematic and alienating. Nevertheless, as I have argued repeatedly in this volume, the continual generation and regeneration of critical perspectives remains vital in the struggle for *better* technoscience and a *better* world.

All of the investigations assembled in the preceding chapters combine historical contextualization with close analysis of specific forms of popular culture. Attention to detail and specificity have been features of both cultural studies and recent science and technology studies. Indeed, since the 1980s many practitioners of science studies and cultural studies have shared a concern with the detail of practices and forms (Latour and Woolgar 1986; Johnson 1983). The articulation of the epistemological, theoretical and methodological implications and consequences of attention to specificity and the resultant challenges to binary constructions of science and society and micro and macro analysis have been a great strength of science studies since the 1980s (Latour and Woolgar 1986; Law and Hassard 1999; Mol 2002). The research presented in this volume has been influenced by these developments. However, cultural studies sensibilities have made me uneasy about versions of science and technology studies in which technicist mappings and more or less abstract descriptions predominate. Power relations, historical patterns and social justice, which effectively instantiate the technoscientific status quo, do not figure in such research.

The tropes which structure this book underscore the multiple and dispersed makings of technoscience in the contemporary Western world. Technoscientific heroes, stories and spectacles are complex cultural productions. Pursuing various manifestations of the making of scientific heroes, the telling of technoscientific stories and the witnessing of technoscientific spectacles carried me away from the canonical sites of science studies scholarly investigation: the laboratories, the clinics and so on. This reorientation enabled me to provide a set of exemplar case studies of the making of technoscience which illustrate that scientists, technicians, clinicians and technologists are not the only agents in the making of technoscience. While much recent science studies has successfully challenged anthropocentric visions of technoscience and emphasized the vast and complex array of actants and materials mobilized in science, its repertoire of human actants and locations is often surprisingly limited.

At the end of the twentieth century public understanding of science caught the attention of many policy makers and an identifiable academic community (Wynne 1995). In the early twenty-first century, 'public engagement with science' has replaced it as a key fulcrum for initiatives and investigations concerned with the social relations of science and technology. However, the terms of reference for such interventions have been restricted (Wynne 1995; Irwin and Wynne 1996). Too often science emerges within such scholarship and policy generation as beaming out from a designated scientific source (a scientist, a text and so on). The public is called upon to respond to this warming light and to bask in the vision(s) it provides, rather than wallowing in ignorance. This only slightly caricatured picture conveys quite a different understanding of science to the one I have presented in this book. I show the continual makings and remakings of science that are part of daily life in the Western world in the late twentieth and early twenty-first centuries. The general public is always and already palpably engaged in science and technology. As the various studies in this volume demonstrate, we are all agents in the making of science and technology, and scientists, technologists and policy makers cannot control or regulate the generation of meanings around it.

The version of cultural studies of science and technology on offer in this volume flies in the face of the contours and divisions within recent social studies of science and technology. In the early 1990s there were a number of striking critical interventions by science studies researchers who deplored the field's failure to address science in popular culture (Hilgartener 1990; Shapin 1990; Cooter and Pumphrey 1994). However, these scholars did not seem to be aware that a body of original research directly relevant to their concerns was assembling in the various manifestations of feminist cultural studies of science and technology traced in Chapter 2 of this volume. The emergence of the public understanding of science as a distinct academic field, affiliated with a science policy agenda, to some extent pre-empted the potential impact of this positive reflective push. Much of this field became narrowly policy focused: oriented towards documenting and addressing levels of 'scientific literacy' and the communication skills of scientists. In the wake of a series of 'crises' in public attitudes to science and technology (in the UK this included CJD (Creutzfeld-Jakob disease), BSE (bovine spongiform encephalopathy), MMR (the measles, mumps and rubella vaccine), the genetically modified (GM) food debate, foot and mouth disease and avian flu), public engagement with science has gradually replaced public understanding of science as the main focus of policy makers. Too often 'public engagement with science' has been regarded as a form of social engineering and experimentation to make contemporary science and technology acceptable to a wide public. As I have demonstrated through my own investigations and with reference to those of others undertaking cultural studies of science, the public is fully engaged with science and technology. What is needed is a more profound and complex understanding of the forms of this engagement. I am suggesting that cultural studies of science and technology offers more promise than either public understanding of science or public engagement with science research as a set of approaches to science and technology in public life.

I hope that this book will convey the challenge and excitement I have experienced as an active participant in the bringing together of science studies, cultural studies and feminist studies. I have indicated some of the promising and distinctive aspects of this encounter. These have included the acknowledgement that technoscience is fully cultural, but also the recognition that the delineation of what this means or involves in any particular instance, site or enactment is an open research question. The rich repertoire of resources that have been brought into this field can be both exhilarating and intimidating. This book has sought to emphasize the insights and perspectives they bring to the social studies of science and technology. My intention has been to diffuse misgivings by showing how these can be and have been mobilized to enhance understandings of our technoscientific lives. Moreover, I have demonstrated the promise of cultural studies of science and technology studies in its capacity to address the many registers of technoscience – emotional as well as cognitive.

Returning to Joni Mitchell's lyrics with which I opened this book (see p. 1), I would insist that I did not wake up one morning to *discover* that I was doing feminist cultural studies of technoscience. Nevertheless, in assembling this book I have come more fully to appreciate why doing feminist cultural studies of technoscience is something I and others want to do.

Notes

1 I woke up one morning and discovered that I was doing feminist cultural studies of science and technology

1 Steven Shapin (1992) calls for more awareness of and reflexivity about the history of the disciplines of the history and sociology of science.
2 For an interesting account of the setting up of 'an educational program of cultural studies of science and technology' at Georgia Institute of Technology, see Balsamo (2000).
3 Rouse (1992) nominates a particular group of contributors to this field, including some researchers who may have been surprised to find themselves identified as doing cultural studies of science.
4 Postcolonial studies, itself an interesting politicized hybrid trans-discipline, has been brought into dialogue with science and technology studies, in the new field of postcolonial technoscience studies (see, for example, W. Anderson 2002; McNeil and Castañeda 2005).
5 Keller refers to 'contemporary studies of science', which she sees coming into being in the last decades of the twentieth century with 'the recognition of the distinction between science and nature' (Keller 1999: 234). While I fully agree that this intellectual marker has been a distinguishing feature of science studies for the last few decades, it remains the case that the institutional and intellectual pedigree of science studies stretches back further. See David Edge's (1995) reflections about the emergence of science and technology studies in the UK from the mid-1960s.
6 The proliferation of cultural studies since the 1980s in a wide range of incarnations and institutional forms raised questions about its political orientation, allegiance and significance. This is the subject of ongoing debates and controversy (McGuigan 1992). For an earlier account of cultural studies, see Johnson 1983.
7 Some important work and key figures in the field of science and technology studies have had strong links to left and alternative political movements and activities including (to mention just a few) the Radical Science movement and the Society for Social Responsibility in Science in the United Kingdom and the United States from the 1970s into the 1990s, the science shops movement in the Netherlands from the 1970s to the 1990s, the ecology movement, and extensive affiliations to trade unions in Europe in the late twentieth century and early twenty-first centuries. The *Radical Science Journal* (vols 1–17), which was published between 1974 and 1985, and *Science for People*, published by the British Society for Social Responsibility in Science came out of these movements, as did the US publication *Science for the People*. For a polemical account of developments within the radical science movement at the end of the 1970s, see H. Rose and S. Rose 1979. Moreover, since the 1980s, various connections have been forged between science studies and explicitly political movements and

campaigns, including the anti-nuclear movement, HIV/AIDS activism and the ecology movement.
8 See, especially, 'Part Three: Science and technology', in Franklin *et al.* 1991b: 127–218; and ch. 7, 'Technologies of the body', in Thornham 2000: 155–83.
9 Haraway (1997: 280, n.1) also mentions Paul Rabinow's discussion of this term and his linking of it to Heidegger's perceptions of technicity: the transformation of the entire world into a set of resources which are to be exploited.

2 Feminist cultural studies of science and technology

1 For a mapping of these, see Lykke 2002, forthcoming.
2 For example, some of the work of Constance Penley, Andrew Ross and Sarah Franklin blends elements of the traditions of cultural anthropology with those of British feminist cultural studies.
3 Although, as Constance Penley and Andrew Ross (1991a) have argued, there are strands of postmodernist theory which are uncritically celebratory in their attitudes towards technology.
4 The list of feminist-influenced science fiction writers of this period is extensive and would include, to name but a few key figures: Margaret Atwood, Marion Zimmer Bradley, Olivia Butler, Suzy McKee Charnas, Zoe Fairbairns, Sally Miller Gearhart, Ursula Le Guin, Anne McCaffrey, Vonda McIntyre, Naomi Mitchison, Marge Piercy, Joanna Russ, Pamela Sargent and James Tiptree Jr.
5 Hilary Rose (1994: 209) nominates, rather than Mary Shelley, Margaret Cavendish, the seventeenth-century English philosopher, duchess and author of the utopia *The Description of the New World Called the Blazing World* (1688), as the original 'foremother' of science fiction.
6 The *Reload* collection is particularly concerned with cyberfiction. The introduction to the volume (Booth and Flanagan 2002) and the chapters by Booth (2002) and Hollinger (2002) provide valuable reviews of and fresh perspectives on cyberfiction and feminist and queer science fiction more generally. Another interesting edited volume (Larbalestier 2006) contains a collection of eleven feminist science fiction stories and individual commentaries on each of them.

3 Newton as national hero

1 See Haynes 1994 (ch. 4), who also reviews some of the representations of Newton in eighteenth-century poetry.
2 Beaven's *Newton's Niece* (1994) imaginatively explores attributions regarding Newton's sexuality, including voyeurism and homosexuality, while foregrounding the scientist's niece, Catherine (Kit). *The Newton Letter* (Banville 1999) is a novel about an historian of science undertaking research on Newton.
3 In fact, it was one of the group Yeo studies, William Whewell, who first coined the term 'scientist', in 1833.
4 On the use of case studies in recent forms of social studies of science, particularly Actor Network Theory, see Erickson 2005: 82–5. He explains that '[t]o achieve its aims, actor-network theory proceeds by identifying case studies of interest, and then investigating the network of relations that emerge from given situations' (Erickson 2005: 82). Both Lauren Berlant (1997) and Lynne Pearce (2004) discuss the shift from textual analysis to 'case studies' in literary-oriented cultural studies work. Berlant notes that the term 'case study' seems more appropriate than 'text' to denote the sorts of analyses she undertakes, which focus on literary and filmic texts, as well as cultural events (see Berlant 1997: 11–15). See also Pearce 2004: 218, fn.6.

5 Jordanova contends that

> These cultural associations between science, medicine and nationhood were forged by two related processes. First, practitioners of science and medicine actively built imagined communities for themselves, which were based, more or less, on national boundaries. ... Second, practitioners identified both themselves as individuals and their communities as collectivities with a relevant nation.
> (Jordanova 1998: 197–8)

4 Making twentieth-century scientific heroes

1 Anna modifies Watson's framework of 'winners' and 'losers' to include 'competitors' (Goodfield 1982: 212). She wonders if he or Crick ever get depressed (Goodfield 1982: 59).
2 For an analysis of the history of cinematic representations of scientists as bad, mad or dangerous, see Frayling 2005.
3 There was some concern that the mores and ethics displayed in Watson's account would cast science in an unfavourable light. See Sayre 1975: esp. 195; Stent 1980b; Yoxen 1985.
4 Through part of the book Maurice Wilkins looms as a rival but he is presented as a rather subdued one who is easily transposed into a collaborator.
5 There is a very brief reference to Watson's meeting with Dorothy Hodgkin, the British crystallographer, who would become a member of the Royal Society (Watson 1969: 54). However, there is no account of her work in the text.
6 It should be noted that both June Goodfield and Evelyn Fox Keller were themselves accomplished scientists. Keller has written about her own life as a scientist in a collection of articles in which feminists reflect about their work lives (Keller 1977).
7 For a discussion of the gender divisions and relations of science during this period, see Rossiter 1982, 1995; Abir-Am and Outram 1989; H. Rose 1994.
8 Sayre does not say much about Franklin's Jewish background. This looms larger in Brenda Maddox's (2002) more recent biography of Franklin.
9 This is not to say that these scientists were not shown as having good collegial relations. Sayre and Brito are portrayed as engaging in extended positive collaborations. While McClintock is shown as a rather lone researcher, her mentoring skills are also demonstrated in Keller's account. In all cases, it is striking that these collaborations are generally with those who are either juniors in the scientific hierarchy and/or social outsiders in the world of science because of their gender or national identities. For example, Anna's collaborators in the United States are almost exclusively foreigners. For a discussion of Goodfield's failure to reflect on this aspect of Brito's career, see Abir-Am 1982: 2.
10 Hubbard complains that 'the book devotes too much time to defending Franklin's "femininity" in the traditional sense of the word, but that is a matter of taste' (Hubbard 1976: 236).
11 For a psychoanalytical perspective on these issues, developed with reference to the history of molecular biology, see Keller 1992.
12 In fact, Evellen Richards and John Schuster (1989a) criticized Keller for being conservative in adhering to a conventional narrative of scientific discovery.
13 McClintock was awarded a Nobel Prize shortly after the publication of Keller's biography.
14 Each of the authors of the biographies employs specific methodological techniques to facilitate her biographical studies. Sayer's access to Franklin's papers and to interviews with Franklin's colleagues, Keller's interviews with the reclusive McClintock and Goodfield's quasi-ethnographic methods, which include observations, interviews, letters and telephone conversations, are all invoked to underscore the accuracy of their pictures of both the scientist and the scientific world she inhabited.

156 *Notes*

15 This was the title of the first edition of this book. The second edition, published in 1986, appeared under the title: *Laboratory Life: the construction of scientific facts.* The authors explained the reasons for this alteration in the 'Postscript to the Second Edition' (Latour and Woolgar 1986: 281).
16 For a list of the main 'laboratory studies' which had appeared up to 1986, see Latour and Woolgar 1986: 285, n.4.
17 Carol Gilligan's (1982) contestation of Lawrence Kohlberg's theory of moral development provided the original framing for the notion of a distinctively feminine 'ethics of care' – the idea that women, in contrast to men, develop a moral sensibility through their investments in relationships. Gilligan's ideas were extremely influential, particularly within feminist scholarship in the late twentieth century. However, there was also notable critical response to Gilligan's work. One nub of controversy revolved around the nature of the bond between women and care. Questions were raised about whether this was an essentialist – a 'natural' – bond, an acquired characteristic or a problematic symptom of female oppression.
18 See Harding's long note of references supporting her claim (Harding 1991: 296, n.1).
19 The wave of laboratory studies in the late 1970s and early 1980s and his own research on Pasteur informed Latour's (1983) declarations that laboratory studies afforded understandings of the social and political specificities of science which challenged the distinction between 'micro' and 'macro', 'internalist' and 'externalist' science studies, and the 'inside' and 'outside' of science. While Latour's insistence that the laboratory was never 'immune from social forces' (Latour 1983: 156) was important, as the title of his 1983 article suggests ('Give me a laboratory and I will raise the world'), his preoccupation with the laboratory as the paramount site of science deflected attention from other locations and agents in the making of science. Latour subsequently raised critical questions about the assumptions which informed such laboratory studies when he commented that 'the main limitation of laboratory studies including my own' was that

> They start out from a place without asking if this place has any relevance at all and without describing how it becomes relevant. In only a very few cases are laboratories the place to start with if we wish to see science in the making.
> (Latour 1988: 261, fn.15).

20 See Keller's (1992) own attempt to pursue a psychosocial analysis of science.
21 Actor Network Theory explicitly addressed the complexity of the making of science. A key element in this was giving more attention to non-human actors. This orientation was linked to scepticism about the attribution of agency to scientific heroes. Latour's study of *The Pasteurization of France* (1988) provides a new architecture for conceptualizing the French national scientific hero which deconstructs Pasteur's agency and heroism. For another study of a more recent scientific hero, Stephen Hawking, see Mialet 1999.
22 Public understanding of science (PUS) is a complex field involving both policy makers and academics, particularly in the United Kingdom and the United States. Brian Wynne notes that '[d]espite the long career of these discourses, it is only since the mid-1980s that the PUS issue took on the trappings of institutionalization' (Wynne 1995: 361). Wynne (1995) offers a critical commentary on the emergence of this field. The journal *Public Understanding of Science*, which was launched in 1992, has been the main academic periodical of this sub-field. See also Gregory and Miller 1998.
23 Haraway, in effect, offers a 'diffraction' (my invocation of her own concept; see Haraway 1997: 14, 272–3) of Shapin and Schaffer's figure of the 'modest witness' through the story of Boyle and the experimental way of life (Haraway 1997: 33).

5 New reproductive technologies

1 As Pat Spallone (1992) and Karen Throsby (2004: ch. 8) argue, there are important connections between developments in these two fields.
2 The first version was given at Allegheny College, Meadville, Pennsylvania, in 1989 (McNeil 1993).
3 For a discussion of how feminist thought is created as well as communicated through rhetorical and other stylistic innovations, see Pearce 2004.
4 Childlessness has almost exclusively negative connotations. There is no noun equivalent to the somewhat more positive adjective, childfree. See Throsby's (2004) discussions of the linguistic constraints in this designation. See also Campbell 1999; Carter and Carter 1998; Ireland 1993; Woollett 1996.
5 In this respect I was influenced by the work of Sandra Harding (1986, 1991), Sharon Traweek (1988) and other feminists who have problematized the position of the neutral observer or analyst.
6 The history of in vitro fertilization (IVF), for example, can be traced back to experiments with mammalian eggs in Vienna in 1878, with the first successful extra-corporeal conception being reported in 1934. In the late 1940s human IVF was realized in the USA, but social pressure forced a halt to such experimentation in the 1950s (Lorber 1988: 119).
7 Objections have been raised about this term on grounds of it having religious connotations.
8 Throsby offers a similar proviso when she explains that she uses the term NRTs 'although this is not to suggest that they constitute a completely new departure or to disconnect them from their own histories' (Throsby 2004: 10).
9 Whereas Charis Thompson (2005: ch. 2) sees a clear break between a phase of feminist critical work on NRTs (from 1984 to 1991) and a later phase in which feminists were more positively oriented towards these technologies, I would highlight the importance of critical feminist perspectives across this entire period.
10 Another reason was the fear of the possible commercial exploitation of less well-off women who might be attracted to offer their services for financial reasons.
11 Amongst the many versions of this story are those offered by Donnison 1977; Ehrenreich and English 1979; Kitzinger 1967; Oakley 1976, 1984; Rich 1977; Stacey 1988: ch. 17.
12 This was the precursor to the UK Human Fertilisation and Embryology Authority (HFEA), which was established in 1990.
13 This redefinition of pregnancy can be linked to a long-term pattern of devaluing women's experientially based knowledge of pregnancy and childbirth. For a discussion of the elimination of 'quickening' as the marker of pregnancy in the Western world, see Duden 1993a: 79–98.
14 Thompson continues:

> The annual meeting of the ASRM [American Society for Reproductive Medicine] displays the extent to which drug companies and instrument makers collaborate with practitioners in research and in many other aspects of American assisted reproductive technologies. The development of recombinant fertility drugs, after an acute shortage of naturally derived ovulation-induction drugs in the mid-1990s, marked a new level of standardization of ovarian stimulation in IVF and illustrates well this aspect of ARTs.
>
> (C. Thompson 2005: 233)

15 Some early attempts to describe the commercial dimensions of NRTs were offered by Mies 1986 and, with particular reference to Australia, by Brown *et al.* 1990 and Koval 1990.

16 Naomi Pfeffer (1992) discusses some of the features of the provision of IVF within the British NHS in the 1980s and early 1990s and makes specific contrasts with practices in other countries.
17 Strathern makes a related, but somewhat different point, when she notes that 'like the Warnock Report (1985) before it, the Glover Report [on Reproductive Technologies to the European Commission, 1989] goes out of its way to comment on commercialisation in transactions involving gametes'. She contends that neither of these reports acknowledged that 'the market analogy has already done its work: we think so freely of the providing and purchasing of goods and services that transactions in gametes [are] already a thought-of-act of commerce' (Strathern 1992b: 37).
18 For critical feminist perspectives on this Royal Commission and on NRTs, see Basen *et al.* 1993, 1994.
19 These narratives could also be associated with Michel Foucault's term 'subjugated knowledges' in that they 'have been disqualified as inadequate ... or insufficiently elaborated: naïve knowledge, located low down the hierarchy beneath the required level of cognition or scientificity' (Gordon 1980: 82).
20 There have been many modifications to and additions to this basic procedure since the late 1970s, which include pre-implantation genetic diagnosis, egg donation, sperm donation, egg freezing, to mention but a few of these.
21 For a fuller account of this procedure, see Crowe 1990b: 29–30; Sherwin 1995: 266–7. For descriptions of women's experiences of IVF, see Duelli Klein 1989; Brown *et al.* 1990: 91–2; Crowe 1990a: 58–66. Koval and Scutt also describe the procedures involved in 'testing for infertility' (Koval and Scutt 1990: 89). Franklin presents the stages in the process of IVF and argues that, with the development of this technology, 'each stage in the sequence is broken down into smaller and smaller stages' (Franklin 1992: 83). Kirejczyk contends that there is a tendency to associate IVF with 'the moment of fertilization only' and to ignore the fact that 'the entire process of female procreation is being affected' (Kirejczyk 1993: 511).
22 See Chapter 6 for a discussion of the limits on the endorsement of such practices and of the controversies which emerged around them in particular circumstances.
23 Meg Stacey makes a related observation when she asks 'why in liberal democracies ... involuntary childlessness' is 'seen as an issue of health and well-being and not a social or specifically a population problem' (Stacey 1992b: 2).
24 The problems of women who do not have children for social reasons are often dismissed or ignored. A further instance can be found in Anne Woollett's (1996) otherwise rather sensitive study of the accounts of childless women and of women with reproductive problems. She refers to women who have: '*chosen* not to have children' (my emphasis) and then observes that some of these 'may not be in a social position to have children' (Woollett 1996: 50). Although it seems that many more women are bringing up children on their own or remaining childless in Britain and North America, this does not obfuscate the difficult social conditions which often shape these choices. See also Ireland 1993.
25 This had already been clearly established in relation to abortion rights in Britain (see Science and Technology Subgroup 1991).
26 The controversy raises further questions about whether such women were subverting conventions (in appropriating reproductive technologies outside heterosexual norms) or conforming to gender norms in becoming patients willing to undergo medical treatment to become mothers.
27 Female (and indeed male) academics in the UK are predominantly white and middle class.
28 Brown, Fielden and Scutt make a similar observation about the lack of concern about minimal healthcare provisions for black Australian women and their children, in contrast to the allocation of resources for NRTs (Brown *et al.* 1990: 102).
29 The acronym VLA refers to the UK Joint Medical Research Council and Royal College of Obstetricians and Gynaecologists' Voluntary Licensing Authority for

Human in Vitro Fertilisation and Embryology, established in 1985 as a result of the recommendations of the Warnock Committee Report (Committee of Inquiry into Human Fertilisation and Embryology 1984). The HFEA is a UK statutory body established by the Human Fertilisation and Embryology Act (1991) to monitor IVF and DI clinics and the storage of eggs, sperm and embryos. See www.hfea.gov.uk/cps/rde/xchg/hfea (accessed 30/11/06).
30 According to the British HFEA website, the typical cost of an IVF cycle in the United Kingdom is £3000, with consultation, drugs, tests and embryo freezing being added to this basic cost (www.hfea.gov.uk (accessed 30/11/06)).
31 Although technological determinism has been a feature of some cyberfeminism. For a critique of these, see Squires 1996; Stabile 1994; see also pp. 141–3.
32 Ann E. Kaplan delineates three 'mother related discourses' that constitute 'representations that appear to have been given a wide circulation in North American culture by a variety of means' (Kaplan 1992: 19–20). She labels these the Rousseauian discourse of the early modern period, Darwinian/Marxist/Freudian discourses of the high modernist period and postmodern discourses.
33 See Leah Wild's diary (2000–1) and discussion in Chapter 6.
34 In spring 1993 British listeners and viewers were offered two fictional serial representations of these issues. Infertility treatment became a narrative strand in the popular BBC Radio 4 soap opera *The Archers*, and a three-part drama series on BBC television (*Me, You and It*) presented a couple undergoing infertility testing and treatment. Since then there have been countless representations of experiences of IVF in Britain. Amongst these are Timberlake Wertenbaker's play *The Break of Day* (1995) and a documentary series – *Making Babies* – broadcast on BBC television in May 1996 based on Robert Winston's IVF clinic in Hammersmith, London. Infertility treatment has figured in many novels and other fiction since the 1990s. Amongst such popular fictional texts are Kathy Lette's *Foetal Attraction* (1993) and Ben Elton's *Inconceivable* (1999). *The Archers* has recently (2006–7) featured another storyline about possible infertility.
35 It is striking that in this film every female character of child-bearing age becomes pregnant. Faludi considers the representation of babies and foetuses in Hollywood films of the 1980s as a crucial part of the backlash against feminism and as issuing a call for 'women's return to total motherhood' (Faludi 1992: esp. 162–6).
36 Tsing (1990) undertook research on public responses to cases in which women were charged with 'perinatal endangerment'. She shows the popular conceptualization of these women as 'anti-mothers'. See also Pollitt 1990. On the extension of maternal responsibilities to 'the moment of conception or even before', see Lupton 1995: 155.
37 Carol A. Stabile observes that class and racial divisions may be crucial in these legal prosecutions. She suggests that, for middle-class women, these are more likely to involve 'court-imposed caesareans' since they have 'access to prenatal care (and the ability to pay for it)', 'while fetal neglect cases will largely be aimed at poor women, many of whom are women of color' (Stabile 1994: 95–6, n.6).
38 Of course, this begs the question of what constitutes 'a full course' of technological treatment, particularly given the recent rapid pace of developments in reproductive technoscience.
39 For an account of leaving IVF, see Murdoch 1990. Charis Thompson reports that, in all of her extensive ethnographic research on IVF clinics, she only ever encountered one woman who withdrew from treatment 'without a recommendation from the physician, without severe financial pressure, and without being pregnant' (C. Thompson 2005: 94).
40 The story of the pregnancy of Patricia Rashbrook, a consulting psychologist who lives in Lewes, England, which involved the implantation of a donor egg in a clinic in Russia by an Italian infertility doctor, broke in the United Kingdom in May 2006. The British press reported the birth of her son in July 2006 (see www.guardian.co.uk/

uk_news/story/0,1815854,00.html) and the *Daily Mail* and *Mail on Sunday* bought 'exclusive rights' to the story of Rashbook and her partner. This involves them recounting their experience of conceptive technology treatment, pregnancy and parenthood in serial instalments.

41 Robert Winston, a leading British IVF specialist, describes IVF as 'physically difficult and very time consuming'. He continues: 'Indeed, the woman usually has to give up work during treatment and, unless she is prepared to travel, may have to live temporarily in digs or a hotel during the treatment cycle so that she may be "on call"' (Winston 1987: 76). See also Crowe 1990a: 63–4.

42 Stabile's argument was made with specific reference to the USA, and she observes that '[a]s long as this specific laborer [the pregnant woman] remains invisible, the discourse of fetal autonomy is going to be difficult to overcome' (Stabile 1994: 94).

43 Nevertheless, there has been extensive media attention given to women ostensibly *avoiding* labour in childbirth, through elective Caesarean births. The widely circulated dismissive 'too Posh to push' label referred to Victoria [Posh] Beckham's decision to pre-arrange a Caesarean birth and portrayed affluent women opting for such elective surgery as rich and lazy.

44 For an account of the filming of Louise Brown's birth, see Challoner 1999: 49; see also Throsby 2004: 1.

45 Deborah Lupton (1995: 140) observed that, in press coverage on breast cancer in Australia between 1987 and 1990 links were drawn between childless women pursuing career interests and susceptibility to this disease. This research and another project on a 'pill scare' in the British press in 1983 (Wellings 1985) show the circulation of the message that 'women who refuse to adopt the traditional feminine maternal role and choose instead a professional, well-paid career are courting disaster and bodily punishment for not fulfilling their biological destinies' (Lupton 1995: 140).

46 Lesley Doyal (1987) showed that ability to pay, as well as geographical region and length of the waiting lists for treatment, is also crucial in Britain. A 1991 survey undertaken by the Canadian Royal Commission on New Reproductive Technologies 'confirmed' (their term) that 'upper middle class, well-educated, married couples are more likely than other people to use IVF treatment'. The survey showed that 80 per cent of the IVF patients interviewed had a family income of over $50,000 (compared to 33.3 per cent of the general population who had such incomes) (Royal Commission on New Reproductive Technologies 1993: 554). Lorber notes that in the USA, 'despite the high rate of infertility in poor and black populations, these groups have had a low rate of IVF use' (Lorber 1988: 119). As this suggests, even within Northern countries relatively privileged women have greater access to infertility treatment. More recent research by Throsby (2004) and C. Thompson (2005) shows that, although access to NRTs has been extended, middle- and upper-class women and couples predominate in the profile of users.

47 Naomi Pfeffer (1992: 48–9) reported that the *Independent* (perhaps the British newspaper most associated with middle-class consumers) produced two editions of its guide *How to Choose a Test-Tube Baby Clinic*. See Whitbeck 1988 on the influence of market models and Rothman 1986 on the 'commodification of life' associated with NRTs. Marilyn Strathern discusses the 'hidden prescription that we *ought* to act by choice' and 'the prescriptive fertility ... that accompanies prescriptive consumerism' (Strathern 1992b: 36) in exploring the relationship between enterprise culture and new reproductive technologies.

48 See also Stacey's comments about the 'high normative value ... placed upon biological procreation' (Stacey 1992b: 37) in the Western world. I have suggested that this is not generalized, but that it is focused on particular groups.

49 See also Sherwin 1995: 276–7. On the recent preoccupation, particularly in the United States, with genetic parenthood in the face of other glaring social problems amongst women and children, see Henifin 1988: 5.

6 Telling tales of reproduction and technoscience

1 In fact, as noted previously, Wild recounts her experience of PGD (pre-implantation genetic diagnosis). Karen Throsby suggests that the use of the IVF label is perhaps significant here because 'existing technologies become normalized in comparison to newly controversial technologies [such as PGD]' in this field of technoscientific medicine and because 'PGD patients can mobilize the relative normality (and social familiarity and intelligibility) of IVF in order to distance themselves from hurtful "designer baby" associations' (Throsby 2004: 192).
2 Plummer subsequently explored the conflicts he considered to be symptomatic of 'intimate citizenship' in a set of lectures, the collected texts of which appeared in *Intimate Citizenship: private decisions and public dialogues* (Plummer 2003).
3 In recent years there has been an extensive debate about biopolitics and bio-citizenship. Foucault's characterization of the modern era as the 'age of biopolitics', which he contended dawned in the eighteenth century, has been an important touchstone for such debates (see, especially, Foucault 1979). Agamben (1998) and Bauman (1989) offer rather different perspectives on modern biopolitics, focusing on its negative enactments, particularly in the twentieth century. Rabinow supplements the concept with his notion of 'biosociality' (Rabinow 1996). Carolos Novas and Nikolas Rose, influenced by Foucault, have forged the concept of 'biological citizenship' with specific reference to developments in the Western world in the late twentieth and early twenty-first centuries (see Novas and Rose 2000; N. Rose 2001). Charis Thompson specifies this concept a bit further in employing the term 'biomedical citizenship' (C. Thompson 2005: 6). See also Petryna 2002.
4 Charis Thompson notes, with particular reference to the United States, 'the parent-centred extension of the idea of reproductive rights' (C. Thompson 2005: 7), which she associates with the development of assisted reproductive technologies in the last decades of the twentieth century and which she contrasts with the linking of rights to parenthood to the interests of the child, particularly in adoption practices.
5 The pattern here is similar to that traced by Scott (1992) in her discussion of the invocation of 'experience'.
6 At the end of the twentieth and beginning of the twenty-first centuries, the topic of infertility and NRTs had become a staple narrative line for radio and TV dramas, drama-documentaries and even comedy series in the UK and North America (e.g. *The Archers, Friends*). It has also featured in a number of popular novels (see Lette 1993; Elton 1999).
7 I use the designation 'lesbian, bisexual and single heterosexual women' as a general designator of women who are not involved in heterosexual relations at the time they seek to reproduce. They may identify as lesbian, bisexual or heterosexual or may not explicitly adopt any label regarding their sexual identity.
8 This term refers to the phenomenon of people travelling to different parts of the world to secure gametes or services that will enable them to reproduce. This practice has proliferated since the 1990s.
9 The most famous British case concerned Diane Blood's appeal to use her dead husband's sperm, which she successfully won in 1997, against the strictures of the Human Fertilisation and Embryology Act (1990), which banned such use if explicit permission had not been documented. Blood subsequently gave birth to two children (in 1999 and 2002) after sperm implantation in a clinic in Brussels. For her account, see Blood 2004.
More recently, Natallie Evans has pursued her case to use the frozen embryos which derived from IVF treatment she underwent during her relationship with Howard Johnston. Ms Evans' relationship with Johnston ended and she subsequently had cancer, which has rendered her infertile. He has refused her permission to use the embryos. The UK courts turned down Ms Evans' case and it was also rejected in the

162 *Notes*

European Union courts in March 2007. (For the judgment on this case, see http://news.bbc.co.uk/1/shared/bsp/hi/pdfs/07_03_06_echr.pdf (accessed 14/05/07)).
10 The most recent publicized case of a postmenopausal pregnancy and birth in Britain was that of Patricia Rashbrook, which first appeared as a story in May 2006. See Chapter 5, note 46; www.guardian.co.uk/uk_news/story/0,1815854,00.html.
11 These involve appeals for the right to PGD to facilitate the generation of a sibling who could be a donor of tissue to be used in medical treatment of a child who is suffering from a life-threatening disease. The most publicized case in Britain was that of Raj and Shahana Hashmi, who applied to the HFEA for permission to use PGD and tissue typing in trying to generate a sibling who might be able to produce appropriate tissue for the treatment of their son Zain, who suffers from beta thalassaemia. Turned down in December 2002, they were granted permission by the Court of Appeal in 2003 and this decision was upheld in 2005 by the House of Lords. However, Michelle and Jayson Whitaker were unsuccessful in their appeal for permission to use PGD in order to produce a sibling in an effort to secure treatment for their son Charlie, who has diamond blackfan anaemia.
12 Charis Thompson refers to the recent phenomenon of grandparents being surrogates as 'the reversal of linear descent' (C. Thompson 2005: 12).
13 The discussion of technoscientific imaginaries has been a significant trope in recent feminist analyses of genetic and reproductive technologies. See, especially, Van Dijck 1998; Franklin 2000.
14 Charis Thompson claims that the making of heroes of reproductive technoscientific medicine has been a particularly strong feature of the pattern of development of this field in Britain (C. Thompson 2005: 209–10). The media coverage of developments around NRTs in the United Kingdom has also involved the identification of 'maverick' scientists in the reproductive field (Haran *et al.* 2007: ch. 4).

7 National and international spectacle

1 This war is sometimes referred to as 'the Persian Gulf War' (C. H. Gray 1997). It designates the conflict between Iraq and a coalition of approximately thirty nations, authorized by the United Nations to liberate Kuwait, which occurred between 16 January and 28 February 1991. The US took the lead in this coalition, with the UK being the most active European participant in the coalition. Prior to this war, the Iran–Iraq War of 1980–8 had been referred to as the 'Gulf War' or 'Persian War'. Although the 1991 conflict is, for this reason, occasionally called 'the Second Gulf War', it is now commonly labelled the First Gulf War (Gulf War I in this chapter).
2 In their reader on *The Social Shaping of Technology* (2nd edn), Donald MacKenzie and Judy Wajcman comment:

> The single most important way that the state has shaped technology has been through its sponsoring of military technology. War and its preparation have probably been on a par with economic considerations as factors in the history of technology. Like international competition, war and the threat of war act coercively to force technological change, with defeat the anticipated punishment for those who are left behind. ... Military interest in new technology has often been crucial in overcoming what might otherwise have been insuperable economic barriers to its development and adoption, and military concerns have often shaped the development pattern and design details of new technologies.
>
> (MacKenzie and Wajcman 1999: 15)

One quarter of this reader addresses military technology (see MacKenzie and Wajcman 1999: 341–440). However, Michael Ignatieff contends that, towards the end of the twentieth century there were signs that the military could be declining in its role

in technological innovation. He sees this as part of a wider shift in which 'the era of military mobilization of the civilian economy is well and truly over' (Ignatieff 2000: 190). Of course, Ignatieff's assessment was registered prior to the US and UK engagement in Iraq which commenced in 2003.
3 Penny Harvey (1995) uses this phrase in her analysis of Expo '92 but I would suggest that it is also an appropriate label for characterizing Gulf War I.
4 The appearance of the *Journal of Visual Culture* in 2002 is one indicator of the growth of this field. This publication is obviously an important, but by no means the only, location for debate about the methods and conceptual approaches to research in this field.
5 On other forms of sentimental nationalism, see Berlant 1997 and Chapter 6 in this volume.
6 For an exploration of the relationship between popular film and US politics in the aftermath of the attacks on the World Trade Center in September 2001, see Weber 2006.
7 One military leader defended this form of ground engagement in a comment which appeared in the *Guardian* (13 September 1991: 24): 'I know burying people like that sounds pretty nasty but it would be even nastier if we had to put our troops in trenches and clean them out with bayonets' (Colonial Lord Maggart, quoted by Patrick Sloyan in Shapiro 1997: 84).
8 Of course, it did not take long for critical assessments to emerge about the notion of 'surgical strikes' which cast doubt on the accuracy of such targeting.
9 Scarry refers to 'death counts' (Scarry 1985: 70) as one of the few registers of bodily injury. In an interesting note, she also discusses Henry Kissinger's use of the phrase the 'so called "kill-ratios"' in his *American Foreign Policy: three essays* (Kissinger 1969: 105). She observes:

> The qualification is an interesting example of the loss of reality of injuring. The term 'kill ratio' is an accurate and literal description of the ongoing comparisons of the body counts on the two sides in any war. To preface the term with the qualification 'so called' seems an apology for the brutality of language, as though the term itself, rather than the phenomenon it literally describes, required perceptual crudity. That is, since killing has disappeared from our understanding of the event, its sudden intrusive appearance in language ('kill-ratios') seems a distasteful barbarity only introduced at the moment of description.
> (Scarry 1985: 337, n.22)

10 The game is a product of the 3DO company which was released in 1999 and was re-released in November 2002. It was still available at the time of writing in September 2007 – now in DVD format. The following description is offered of the game:

> Gulf War Operation Desert Hammer is a classic arcade-action shooter; a patriotic game of 3D tank combat set in the present day (2001) Persian Gulf. As commander of a prototype M12 tank, nicknamed 'The Hammer', players use an array of weapons and the aid of additional military units to 'go back and finish what we started with Desert Storm.' Gulf War Operation Desert Hammer offers real-world locations and missions with ease-of-use, arcade-style, single and multi-player gameplay.
> (http://uk.gamespot.com/pc/action/gulfwaroperationdh/index.html (accessed 06/06/07))

For an account of the re-release of the game and of its appeal, see David Lazarus's (2003) article in the *San Francisco Chronicle* (www.sfgate.com/cgi-bin/article.cgi?file=/chronicle/archive/2003/03/26/BU2781 (accessed 07/06/07)).
11 Pat Barker's (1996) trilogy of novels, comprising *Regeneration* (originally published in 1992), *The Eye in the Door* (originally published in 1995) and *The Ghost Road* (published

164 *Notes*

in 1996), indicates the drug treatments which were part of the therapeutic support for soldiers during World War I.
12 This assessment would undoubtedly not stand in the early twenty-first century, in the wake of media coverage of the attack on the World Trade Center in September 2001.
13 The film is based on the bestselling memoir of a marine who was in active service during the conflict. Anthony Swofford, *Jarhead: a marine's chronicle of the Gulf War and other battles* (2003).

8 Techno-tourism in Florida

1 I am conscious of the linguistic appropriation implicit in the use of the adjective 'American' to designate the United States, rather than the many nations and peoples who could claim that label. However, for the purposes of this chapter I deliberately invoke the narrow, conventional and rather problematic use of the term.
2 The National Aeronautics and Space Agency was established by President Eisenhower as a civilian agency of the US government on 1 October 1958 'to carry out peaceful exploration and use of space' (NASA 1992: 2). On the relationship between the popular television (and then film) series *Star Trek* and NASA, see Penley 1997. On the gender relations of the space race, with particular reference to the selection and training of women astronauts, see Kevles 2006.
3 Green berets were the markers of the members of US special forces during this period and they became associated with patriotic heroism during the Vietnam War. The 1968 film *The Green Berets* (directed by Ray Kellogg), starring John Wayne, amplified this image in a vehemently right-wing, pro-US government portrayal of the Vietnam War. *Platoon* (1986, directed by Oliver Stone) was offered as an alternative, ostensibly 'more realistic' representation of that conflict.
4 The US-manned space flights through the NASA programme were organized into three sequential series of 'missions' from 1958 to 1981: Mercury, Gemini and Apollo. It was the Apollo mission which landed men on the moon for the first time in 1969.
5 The *Challenger* space shuttle vehicle exploded after launch from Cape Canaveral on 28 January 1986 with seven astronauts on board. Amongst the astronauts killed was Christa McAuliffe, a school teacher from New Hampshire, chosen for the mission as a 'citizen astronaut', who became a widely mourned popular national heroine. For an analysis of the *Challenger* disaster and of McAuliffe's significance in 1990s US popular culture, see Penley 1997: 22–88. For a critical account of NASA's practices and policies in relation to the *Challenger* disaster, see Vaughan 1996. A commission headed by the former attorney general William P. Rogers, subsequently referred to as the Rogers Commission, was established by President Reagan to investigate this disaster. It reported in July 1986, identifying a number of faults in NASA decision-making and other factors which had contributed to the explosion. Bettyan Holtzman Kevles claims that '[t]he Challenger disaster brought the human space program to an abrupt halt' (Kevles 2006: 115). However, she contends that, in the longer term, this disaster served to renew commitment to NASA's space shuttle programme (Kevles 2006: 117–18).
6 The model community that Walt Disney had proposed was subsequently established on the Orlando site. It is called 'Celebration' (Wasko 2001: 23–4). For background and a commentary on this Disney residential site, see Ross 2000; Wasko 2001: 178–82.
7 The wording here is deliberately ambiguous in its reference both to the site, which was realized 'through the sponsorship of corporate America', and to 'the vision of progress' on offer, which was represented as the product of corporate America (Bryman 1995: 106, 130, 145–60; Wasko 2001: 34–5; 158–61). In 1994 the EPCOT Center was renamed EPCOT 94 and it has undergone a series of changes of name (e.g. EPCOT 95, etc.), together with alterations, renovations and some variation in corporate sponsors since then.

8 William F. Van Wert describes the World Showcase as 'a world zoo of different cultures' (Van Wert 1995–6: 200).
9 As Van Wert suggests, EPCOT displays 'both the inadequacies of the past and the large corporation that is already seeing to our needs in the future. ... The House of Energy ends in Exxon. ... The Story of the Land and Harvest comes to us courtesy of Kraft' (Van Wert 1995–6: 203). Many of the corporate sponsors of exhibits at Future World (e.g. Bell Telephone, Exxon, General Motors, Eastman Kodak and Kraft Food Products) had been sponsors of pavilions at the 1939 World's Fair at Flushing Meadows, New York (Zukin 1991: 227).
10 In a recent BBC 4 Radio programme, *Bird in the Air Pump*, part of *The Century that Made Us* series, which was broadcasted on 21 and 25 June 2006, the presenter Ben Woolley explicitly linked this painting with an experiment Robert Boyle's conducted in 1659. See www.bbc.co.uk/bbcfour/documentaries/features/bird-air-pump.shtml (accessed 10/12/2006).
11 This insight sustains the critical use of the term technoscience. See pp. 98–100.
12 See the quoted excerpt from Keller's comments about this issue on p. 59.
13 Annemarie Mol describes Harding's and related strategies in strikingly visual imagery, associating these 'with hopes, for instance, that if the white male gaze is joined by female and colored optics, unbiased knowledge becomes possible, and objectivity is reached after all' (Mol 2002: 154).
14 Martin (1998) draws explicitly on Haraway's (1994) exploration of critical science studies through the employment of the metaphor of the game of cat's cradle.
15 Haraway uses the notion of 'yearning' (Haraway 1997: esp. 127–9, 211–12), explicitly borrowing from bell hooks' evocative use of the term in *Yearning* (hooks 1990).
16 See note 7.

9 Conclusion

1 For an interesting sample of distinctively presented science studies texts, see Latour and Woolgar [1979] 1986; Latour 1987; Law 2002; Haraway 1997; Mol 2002.

Bibliography

Abir-Am, P. G. (1982) 'An alternative model of scientific behaviour? A review of *An Imagined World: a story of scientific discovery*', *Women's Studies International Forum*, 38 (5): 503–7.
Abir-Am, P. G. and Outram, D. (eds) (1989) *Uneasy Careers and Intimate Lives: women in science, 1789–1979*, with foreword by M. W. Rossiter, New Brunswick, NJ: Rutgers University Press.
Adam, A. (1997) 'What should we do with cyberfeminism?', in R. Lander and A. Adam (eds) *Women into Computing: progress from where to what?*, Exeter: Intellect.
Adorno, T. W. (1991) *The Culture Industry: selected essays on mass culture*, edited with intro. by J. M. Bernstein, London: Routledge.
Agamben, G. (1998) *Homo Sacer: sovereign power and base life*, trans. D. Heller-Roazen, Stanford, CA: Stanford University Press.
Akenside, M. (1744) *The Pleasures of the Imagination: a poem, in three books*, London: R. Dodsley.
Althusser, L. (1970) 'Ideological state apparatuses', in *Essays on Lenin and Philosophy*, London: New Left Books.
Amis, K. ([1954] 1968) *Lucky Jim: a novel*, London: Victor Gollancz Ltd.
Amster, H. (1976) 'The double deceit of the DNA – Rosalind Franklin and DNA' (review of Anne Sayre, *Rosalind Franklin and DNA*), *Psychology of Women Quarterly*, 1 (2) (Winter): 200–3.
Anderson, B. (1983) *Imagined Communities: reflections on the origin and spread of nationalism*, London: Verso.
Anderson, P. (1969) 'Components of the national culture', in A. Cockburn and R. Blackburn (eds) *Student Power: problems, diagnosis, action*, Harmondsworth: Penguin Books.
Anderson, W. (ed.) (2002) *Social Studies of Science*, 32 (5–6), special issue on postcolonial technoscience.
Appadurai, A. (1990) 'Disjuncture and difference in the global cultural economy', *Public Culture* 2 (2) (Spring).
—— (1997) *Modernity at Large: cultural dimensions of globalization*, Minneapolis: Minnesota University Press.
Arditti, R., Duelli Klein, R. and Minden, S. (1984) 'Introduction', *Test-Tube Women: what the future holds for motherhood?*, London: Pandora.
Armitt, L. (ed.) (1990) *Where No Man Has Gone Before*, London and New York: Routledge.
Arney, W. R. and Neill, J. (1982) 'The location of pain in childbirth: natural childbirth and the transformation of obstetrics', *Sociology of Health and Illness*, 4 (1): 1–24.
Aronowitz, S. (1993) *Roll Over Beethoven: cultural strife*, London: Routledge.
Aronowitz, S., Martinsons, B. and Menser, M. (eds) (1996) *Technoscience and Cybercutlture*, New York and London: Routledge.

Atwood, M. ([1985] 1987) *The Handmaid's Tale*, London: Virago.
Ault, D. D. (1974) *Visionary Physics: Blake's response to Newton*, Chicago: Chicago University Press.
Balsamo, A. (1999) *Technologies of the Gendered Body: reading cyborg women*, Durham, NC: Duke University Press.
—— (2000) 'Engineering cultural studies: the postdisciplinary adventures of mindplayers, fools and others', in R. Reid and S. Traweek (eds) *Doing Science + Culture*, New York: Routledge.
Banville, J. (1999) *The Newton Letter*, London: Picador.
Barker, M. (1981) *The New Racism: conservatism and the ideology of the tribe*, London: Junction Books.
Barker, P. (1996) *The Regeneration Trilogy* (incorporating, *Regeneration* (1992), *The Eye in the Door* (1995) and *The Ghost Road* (1996)), London: Viking.
Barr, M. (ed.) (1981) *Future Females: a critical anthology*, Ohio: Bowling Green.
Barthes, R. ([1957] 1973) *Mythologies*, selected and translated by A. Lavers, London: Paladin.
Basen, G., Eichler, M. and Lippman, A. (1993) *Misconceptions: the social construction of choice and the new reproductive and genetic technologies*, vol. I, Hull, Canada: Voyageur Publishing.
—— (1994) *Misconceptions: the social construction of choice and the new reproductive and genetic technologies*, vol. II, Hull, Canada: Voyageur Publishing.
Baudrillard, J. ([1991] 1995) *The Gulf War Did Not Take Place*, Sydney: Power Publications.
Bauman, Z. (1989) *Modernity and the Holocaust*, Cambridge: Polity.
Beaven, D. (1994) *Newton's Niece*, London: Faber and Faber.
Beck, U. (1992) *Risk Society: towards a new modernity*, trans. Mark Ritter, London: Sage.
Beer, G. ([1983] 2000) *Darwin's Plots: evolutionary narrative in Darwin, George Eliot and nineteenth-century fiction*, 2nd edn, Cambridge: Cambridge University Press.
Berger, J. (1972) *Ways of Seeing*, Harmondsworth: Penguin.
Berlant, L. (1997) *The Queen of America Goes to Washington City: essays on sex and citizenship*, Durham, NC, and London: Duke University Press.
—— (2000) 'The subject of true feeling: pain, privacy, and politics', in S. Ahmed, J. Kilby, C. Lury, M. McNeil and B. Skeggs (eds) *Transformations: thinking through feminism*, London: Routledge.
—— (2001) 'Trauma and ineloquence', *Cultural Values*, 5 (1) (January): 41–58.
Bewley, S., Braude, P. and Davies, M. (2005) 'Which career first? The most secure age for childbearing remains 20–35', *British Medical Journal*, 331 (17 September): 588–9; available at www.guardian.co.uk/medicine/story/0,11381,1571409,00.html.
Biggs, S. J. (1989) *The Infertility Handbook*, Stroud: Fertility Services Management.
Billig, M. (1994) 'Sod Baudrillard! Or ideology critique in Disney World', in H. W. Simons and M. Billig (eds) *After Postmodernism: reconstructing ideology critique*, London: Sage.
—— (1995) *Banal Nationalism*, London: Sage.
Blake, W. (1972) *William Blake: complete writings with variant readings*, ed. G. Keynes, Oxford: Oxford University Press.
Blood, D. (2004) *Flesh and Blood: the human story behind the headlines*, Edinburgh: Mainstream Publishing.
Booth, A. (2002) 'Women's cyberfiction: an introduction', in Flanagan, M. and Booth, A. (eds) *Reload: rethinking women + cyberculture*, Cambridge, MA: MIT Press.
Bragg, M., with Gardiner, R. (1998) *On Giants' Shoulders: great scientists and their discoveries from Archimedes to DNA*, London: Hodden & Stoughton.
Brooke, H. (1778) *Collection of the Pieces*, London: published by the author, and sold by Mr White, Mr Cadell, Messrs Dilly and Mr Wallis.

Brown, M., Fielden, K. and Scutt, J. A. (1990) 'New frontiers or old recyled? New reproductive technologies as primary industry', in J. A. Scutt (ed.) *Baby Machine: reproductive technology and the commercialisation of motherhood*, London: The Merlin Press.

Bruno, G. (1992) 'Spectorial embodiments: anatomies of the visible and the female bodyscape', *camera obscura*, 28: 239–61.

—— (1993) *Streetwalking on a Ruined Map: cultural theory and the city films of Elivira Notari*, Princeton: Princeton University Press.

Bryld, M. (2001) 'The infertility clinic and the birth of the lesbian: the political debate on assisted reproduction in Denmark', *European Journal of Women's Studies*, 8 (3): 299–312.

Bryman, A. (1995) *Disney and His Worlds*, London and New York: Routledge.

Campbell, A. (1999) *Childfree and Sterilized: women's decisions and medical responses*, London: Cassell.

Cannell, F. (1990) 'Concepts of parenthood: the Warnock Report, the Gillick debate, and modern myths', *American Ethnologist* 17 (November): 667–86.

Carpenter, L. (2006) 'He'd like to have your babies', *Observer Woman*, 8 (August): 36–41.

Carter, J. W. and Carter, M. (1998) *Sweet Grapes: how to stop being infertile and start living again*, revised edn, Indianapolis: Perspective Press.

Cartwright, L. (1992) 'Women, X-rays, and the public culture of prophylactic imaging', *camera obscura*, 29: 19–54.

—— (1995) *Screening the Body: tracing medicine's visual culture*, Minneapolis: University of Michigan Press.

Challoner, J. (1999) *The Baby Makers: the history of artificial conception*, London: Channel 4 Books.

Chartier, R. (1988) *Cultural History: between practices and representations*, trans. L. G. Cochrane, Cambridge: Polity with Blackwell.

Clifford, J. and Marcus, G. (eds) (1986) *Writing Culture: the poetics and politics of ethnography*, Berkeley: University of California Press.

Comfort, N. C. (2001) *The Tangled Field: Barbara McClintock's search for the patterns of genetic control*, Cambridge, MA: Harvard University Press.

Committee of Inquiry into Human Fertilisation and Embryology (1984) *Report of the Committee of Inquiry into Human Fertilisation and Embryology*, London: HMSO.

Cooter, R. and Pumphrey, S. (1994) 'Separate spheres and public places: reflection on the history of science popularization and science in popular culture', *History of Science*, 32, pt. 3, no. 97 (September): 237–67.

Corea, G. and Ince, S. (1987) 'Report of a survey of IVF clinics in the USA', in P. Spallone and D. L. Steinberg (eds) *Made to Order: the myth of reproductive and genetic progress*, London: Hutchinson.

Crick, F. (1974) 'Foreward' to R. C. Oldby, *The Path to the Double Helix*, London: Macmillan.

Crowe, C. (1990a) 'Bearing the consequences – women experiencing IVF', in J. A. Scutt (ed.) *Baby Machine: reproductive technology and the commercialisation of motherhood*, London: The Merlin Press.

—— (1990b) 'Whose mind over whose matter? Women, in vitro fertilisation and the development of scientific knowledge', in M. McNeil, I. Varcoe and S. Yearley (eds) *The New Reproductive Technologies*, Basingstoke: Macmillan.

Cussins, C. (1996) 'Ontological choreography: agency through objectification in infertility clinics', *Social Studies of Science*, 26: 575–610.

Darwin, E. (1799) *The Botanic Garden: a poem in two parts: Part I: The Economy of Vegetation; Part II: The Loves of the Plants* (the 4th edn of Part I and the 5th edn of Part II), London: J. Johnson.

—— (1803) *The Temple of Nature; or, The Origin of Society*, London: J. Johnson.
Davis-Floyd, R. and Dumit, J. (eds) (1998) *Cyborg Babies: from techno-sex to techno-tots*, London and New York: Routledge.
Dawson, G. (1994) *Soldier Heroes: British adventure, empire and the imagining of masculinities*, London and New York: Routledge.
Delamont, S. (1987) 'Three blind spots? A comment on the sociology of science by a puzzled outsider', *Social Studies of Science*, 17: 163–70.
—— (2003) 'Rosalind Franklin and Lucky Jim: misogyny in the two cultures', *Social Studies of Science*, 33 (2) (April): 315–22.
—— (2005) 'Lives of the great women scientists: the never ending story', *Social Studies of Science*, 35 (3) (June): 491–6.
Desmond, A. J. (1984) *The Politics of Evolution: morphology, medicine and reform in radical London*, Chicago: University of Chicago Press.
Desmond, A. J. and Moore, J. (1991) *Darwin*, London: Michael Joseph.
Donawerth, J. (1997) *Frankenstein's Daughters: women writing science fiction*, New York: Syracuse University Press.
Donawerth, J. and Kolmerten, C. A. (eds) (1994) *Utopian and science fiction by women: worlds of difference*, Liverpool: Liverpool University Press.
Donnison, J. (1977) *Midwives and medical men: a history of professional rivalries and women's rights*, London: Heinemann Educational.
Doyal, L. (1987) 'Infertility – a life sentence? Women and the National Health Service', in M. Stanworth (ed.) *Reproductive Technologies: gender, motherhood and medicine*, Cambridge: Polity Press.
Duden, B. (1993a) *Disembodying Women: perspectives on pregnancy and the unborn*, trans. L. Hoinacki, (originally published as *Der Frauenleib als öffentlicher Ort: vom Missbrauch des Begriffs Leben*, 1991), Cambridge, MA: Harvard University Press.
—— (1993b) 'Visualizing "Life"', *Science as Culture*, 3(4), no. 17: 562–600.
Duelli Klein, R. (1989) *Infertility: women speak out about their experiences of reproductive medicine*, London: Pandora Press.
Dyer, C. (2001) 'How can you penalise my son for what I did?', *Guardian* (24 April): (G2 section) 10–11.
Easlea, B. (1980a) *Science and Sexual Oppression: patriarchy's confrontation with woman and nature*, London: Weidenfeld & Nicolson.
—— (1980b) *Witch-hunting, Magic and the New Philosophy*, Brighton: Harvester.
—— (1983) *Fathering the Unthinkable: masculinity, scientists and the nuclear arms race*, London: Pluto.
Edwards, R. G., Steptoe, P. G. and Purdy, J. M (1970) 'Fertilisation and cleavage *in vitro* of preovulatory mature oocytes', *Nature*, 22: 1,307–9.
Edge, D. (1995) 'Reinventing the wheel', in S. Jasanoff, G. E. Markle, J. C. Petersen and T. Pinch (eds) *Handbook of Science and Technology Studies*, Thousand Oaks, CA: Sage.
Ehrenreich, B. and English, D. (1973a) *Complaints and Disorders: the sexual politics of sickness*, Old Westbury, NY: Feminist Press.
—— (1973b) *Witches, Midwives and Nurses: a history of women healers*, Old Westbury, NY: Feminist Press.
—— (1979) *For Her Own Good: 150 years of the experts' advice to women*, London: Pluto Press.
Elton, B. (1999) *Inconceivable*, London: Black Swan.
Enloe, C. (1983) *Does Khaki Become You?*, London: Pluto Press.
Erickson, M. (2005) *Science, Culture and Society: understanding science in the 21st century*, Cambridge: Polity.

Estling, R. (1991) 'Forum: Behold, a virgin shall conceive', *New Scientist* 132 (6) (April): 50–1.
Evans, M. (1998) *Missing Persons: the impossibility of auto/biography*, London: Routledge.
Faludi, S. (1992) *Backlash: the undeclared war against women*, London: Chatto & Windus.
Fara, P. (2000a) 'Faces of genius: images of Newton in eighteenth-century England', in G. Cubitt and A. Warren (eds) *Heroic Reputations and Exemplary Lives*, Manchester: Manchester University Press.
—— (2000b) 'Isaac Newton lived here: sites of memory and scientific heritage', *British Journal for the History of Science*, 33: 407–26.
—— ([2002] 2003) *Newton: the making of genius*, London: Picador.
Farquhar, D. (1999) 'Gamete traffic/pedestrian crossings', in E. A. Kaplan and S. Squier (eds) *Playing Dolly: technocultural formations, fantasies, & fictions of assisted reproduction*, London: Rutgers University Press.
Fauvel, J., Flood, R., Shortland, M. and Wilson, R. (1988) *Let Newton Be! A new perspective on his life and works*, Oxford: Oxford University Press.
Fédération des Centres d'Etude et de Conservation du Sperme Humain, Schwartz, D. and Mayaux, M. J. (1982) 'Female fecundity as a function of age', *New England Journal of Medicine*, 306 (18 February): 404–6.
Firestone, S. ([1970] 1979) *The Dialectic of Sex: the case for feminist revolution*, with intro. by R. Delmar, London: The Women's Press.
Flanagan, M. and Booth, A. (eds) (2002) *Reload: rethinking women + cyberculture*, Cambridge, MA: MIT Press.
Fletcher, R. (2006) 'Reproductive consumption', *Feminist Theory*, 7 (1) (April): 27–47.
Foucault, M. (1979) *The History of Sexuality: Volume I: an introduction*, trans. Robert Hurley (*La Volonté de savoir*), London: Allen Lane.
Franklin, S. (1988) 'Life story: the gene as fetish object on TV', *Science as Culture*, 3: 92–100.
—— (1990) 'Deconstructing "desperateness": the social construction of infertility in popular representations of new reproductive technologies', in M. McNeil, I. Varcoe and S. Yearley (eds) *New Reproductive Technologies*, London: Macmillan.
—— (1992) 'Making sense of missed conceptions: anthropological perspectives on unexplained infertility', in M. Stacey (ed.) *Changing Human Reproduction*, London, Newbury Park and Delhi: Sage.
—— (1993) 'Postmodern procreation: representing reproductive practice', *Science as Culture*, 3 (4), no. 17: 522–61.
—— (1995) 'Romancing the helix: nature and scientific discovery', in L. Pearce and J. Stacey (eds) *Romance Revisited*, London: Lawrence & Wishart.
—— (1997) *Embodied Progress: a cultural account of assisted reproduction*, London and New York: Routledge.
—— (2000) 'Life itself: global nature and the genetic imaginary', in S. Franklin, C. Lury and J. Stacey (eds) *Global Nature, Global Culture*, London: Sage.
Franklin, S., Lury, C. and Stacey, J. (1991a) 'Introduction 1: Feminism and cultural studies: pasts, presents, futures', in S. Franklin, C. Lury and J. Stacey (eds) *Off-Centre: feminism and cultural studies*, London: HarperCollins.
—— (eds) (1991b) *Off-Centre: Feminism and Cultural Studies*, London: HarperCollins.
—— (2000a) 'Introduction', *Global Nature, Global Culture*, London: Sage.
—— (2000b) *Global Nature, Global Culture*, London: Sage.
Frayling, C. (2005) *Mad, Bad and Dangerous? The scientist and the cinema*, London: Reaktion Books.
Fromm, E. (1962) *The Art of Loving*, London: Allen and Unwin.

Gilligan, C. (1982) *In a Different Voice: psychological theory and women's development*, Cambridge, MA: Harvard University Press.

Ginsburg, F. D. and Rapp, R. (eds) (1995) *Conceiving the New World Order: the global politics of reproduction*, Berkeley: University of California Press.

Goodfield, J. ([1981] 1982) *An Imagined World: a story of scientific discovery*, Harmondsworth: Penguin Books.

Gordon, C. (ed.) (1980) *Michel Foucault: Power/Knowledge: selected interviews and other writings, 1972–1977*, New York: Pantheon.

Gray, A. (2003) *Research Practice for Cultural Studies: ethnographic methods and lived culture*, London: Sage.

Gray, C. H. (1989) 'The cyborg soldier: the US military and the post-modern warrior', in L. Levidow and K. Robbins (eds) *Cyborg Worlds: the military information society*, London: Free Association Books.

—— (1997) *Postmodern War: the new politics of conflict*, London: Routledge.

Gregory, J. and Miller, S. (1998) *Science in Public: communication, culture, and credibility*, New York and London: Plenum Trade.

Griffin, S. (1984) *Woman and Nature: the roaring inside her*, London: The Woman's Press.

Grobicki, A. (1987) 'Barbara McClintock: what price objectivity?', in M. McNeil (ed.) *Gender and Expertise*, London: Free Association Books.

Grossberg, L., Nelson, C. and Treichler, P. (eds) (1992) *Cultural Studies*, New York and London: Routledge.

Guerlac, H. (1979) 'Some areas for further Newtonian studies', *History of Science*, XVII: 75–101.

Gunning, J. (1990) *Human IVF, Embryo Research, Fetal Tissue for Research and Treatment, and Abortion: international information*, London: HMSO.

Gusterson, H. (1991) 'Nuclear war, the Gulf War, and the disappearing body', *Journal of Urban and Cultural Studies*, 2 (1): 45–55.

Haimes, E. (1992) 'Gamete donation and the social management of genetic origins', in M. Stacey (ed.) *Changing Human Reproduction*, London: Sage.

Haran, J., Kitzinger, J., McNeil, M. and O'Riordan, K. (2007) *Human Cloning in the Media: from science fiction to science practice*, London: Routledge.

Haraway, Donna (1989) *Primate Visions: gender, race and nature in the world of modern science*, New York and London: Routledge.

—— ([1985] 1991a) 'A cyborg manifesto: science, technology, and socialist-feminism in the late twentieth century', in *Simians, Cyborgs, and Women: the reinvention of nature*, London: Free Association Books.

—— (1991c) 'Situated knowledges: the science question in feminism and the privilege of partial perspective', in *Simians, Cyborgs, and Women: the reinvention of nature*, London: Free Association Books.

—— (1992) 'The promises of monsters: a regenerative politics for inappropriate/d others', in L. Grossberg, C. Nelson and P. Treichler (eds) *Cultural Studies*, London: Routledge.

—— (2003) *The Companion Species Manifesto: dogs, people and significant others*, Chicago: Princkly Paradigm Press.

—— (1994) 'A game of cat's cradle: science studies, feminist theory, cultural studies', *Configurations*, 2 (1): 59–71.

—— (1997) *Modest_Witness@Second_Millenium.FemaleMan$^©$_Meets_OncoMouse^{TM}: feminism and technoscience*, New York and London: Routledge.

—— (2000) *How Like a Leaf: an interview with Thyrza Nichols Goodeve*, London and New York: Routledge.

Harding, S. (1986) *The Science Question in Feminism*, Milton Keynes: Open University Press.
—— (1991) *Whose Science? Whose Knowledge? Thinking from women's lives*, Milton Keynes: Open University Press.
Harvey, P. (1995) 'Nations on display: technology and culture in Expo '92', *Science as Culture*, 5 (1), no. 22: 85–105.
Haynes, R. D. (1994) *From Faust to Strangelove: representations of the scientist in Western literature*, Baltimore, MD: Johns Hopkins University Press.
Henifin, M. S. (1988) 'Introduction: women's health and the new reproductive technologies', in E. H. Baruch, A. F. D'Adamo, Jr. and J. Seager (eds) *Embryos, Ethics and Women's Rights: exploring the new reproductive technologies*, New York: Harrington Park Press.
Hewison, R. (1986) *Too Much: art and society in the sixties 1960–75*, London: Methuen.
—— (1987) *The Heritage industry: Britain in a climate of decline*, London: Methuen.
Hilgartner, S. (1990) 'The dominant view of popularization: conceptual problems, political uses', *Social Studies of Science*, 20: 519–39.
Hird, M. (2004) *Sex, Gender and Science*, London: Palgrave Macmillan.
Hollinger, V. (2002) '(Re)reading queerly: science fiction, feminism, and the defamiliarization of gender (criticism)', in M. Flanagan and A. Booth (eds) *Reload: rethinking women + cyberculture*, Cambridge, MA: MIT Press.
hooks, b. (1990) *Yearning: race, gender, and cultural politics*, Boston: Southend Press.
Hoskins, A. (2004) *Television War: from Vietnam to Iraq*, London: Continuum.
Hubbard, Ruth (1976) 'Review of *Rosalind Franklin and DNA*. By Anne Sayre', *Signs* (Autumn), 2 (1): 229–37.
—— (1990) *The Politics of Women's Biology*, New Brunswick, NJ, and London: Routledge.
—— (2003) 'Review of *Rosalind Franklin: The Dark Lady of DNA*. By Brenda Maddox', *Signs* (Spring): 973–5.
Hudson, M. and Stanier, J. (1997) *War and the Media*, Stroud: Sutton Publishing.
Ignatieff, M. (2000) *Virtual War: Kosovo and beyond*, London: Chatto & Windus.
Ireland, M. S. (1993) *Reconceiving Women: separating motherhood from female identity*, London: The Guilford Press.
Irwin, A. and Wynne, B. (eds) (1996) *Misunderstanding science? The public construction of science and technology*, Cambridge: Cambridge University Press.
Jacobus, M., Keller, E. F. and Shuttleworth, S. (1990) *Body/Politics: women and the discourses of science*, New York: Routledge.
Jasanoff, S. (2005) *Designs on Nature: science and democracy in Europe and the United States*, Princeton: Princeton University Press.
Johnson, R. (1983) 'What is cultural studies anyway?', *Anglistica*, 26 (1–2): 1–75.
Johnson, R., Chambers, D., Raghuram, P. and Ticknell, E. (2004) *The Practice of Cultural Studies*, London: Sage.
Jordanova, L. (ed.) (1986) *The Languages of Nature: critical essays on science and literature*, London: Free Association Books.
—— (1989) *Sexual Visions: images of gender in science and medicine between the eighteenth and twentieth centuries*, Hemel Hempstead: Harvester Wheatsheaf.
—— (1998) 'Science and nationhood: cultures of imagined communities', in G. Cubitt (ed.) *Imagining Nations*, Manchester: Manchester University Press.
Kaplan, E. A. (1992) *Motherhood and Representation: the mother in popular culture and melodrama*, London and New York: Routledge.
—— (1999) 'The politics of surrogacy narratives: 1980s paradigms and their legacies in the 1990s', in E. A. Kaplan and S. Squier (eds) *Playing Dolly: technocultural formations*,

fantasies, & fictions of assisted reproduction, New Brunswick, NJ: Rutgers University Press.

Kaplan, E. A. and Squier, S. (1999) 'Introduction', in E. A. Kaplan and S. Squier (eds) *Playing Dolly: technocultural formations, fantasies, & fictions of assisted reproduction*, New Brunswick, NJ: Rutgers University Press.

Keller, E. F. (1977) 'The anomaly of a woman in physics', in S. Ruddick and P. Daniels (eds) *Working it Out*, New York: Pantheon.

—— (1981) 'June Goodfield, *An Imagined World: A Story of Scientific Discovery*', *New York Times Book Review* (19 April): 8.

—— (1983) *A Feeling for the Organism: the life and work of Barbara McClintock*, New York: W. H. Freeman and Company.

—— (1985) *Reflections on Gender and Science*, New Haven, CT, and London: Yale University Press.

—— ([1987] 1999) 'The gender/science system: or, is sex to gender as nature is to science?', in M. Biagoli (ed.) *The Science Studies Reader*, New York and London: Routledge.

—— (1989) 'Just what *is* so difficult about the concept of gender as a social category? (Response to Richards and Schuster)', *Social Studies of Science*, 19: 721–4.

—— (1992) *Secrets of Life/Secrets of Death: essays on language, gender and science*, New York and London: Routledge.

Kellner, D. (1992) *The Persian Gulf TV War*, Boulder, CO: Westview.

—— (1995) *Media Culture: cultural studies, identity and politics between the modern and the postmodern*, London and New York: Routledge.

Kember, S. (2003) *Cyberfeminism and Artificial Life*, London and New York: Routledge.

Kerouac, J. (1957) *On the Road*, New York: Viking Press.

Kevles, B. H. (2006) *Almost Heaven: the story of women in space*, Cambridge, MA, and London: MIT Press.

Kirejczyk, M. (1993) 'Shifting the burden onto women: the gender character of in vitro fertilization', *Science as Culture*, 3 (4), no. 17: 507–21.

Kissinger, H. (1969) *American Foreign Policy: three essays*, New York: W. W. Norton & Co.

Kitzinger, S. (1967) *The Experience of Childbirth*, revised edn, Harmondsworth: Penguin.

Knight, R. P. (1796) *The Progress of Civil Society: a didactic poem, in six books*, London: printed by W. Bulmer and Co. for G. Nicol.

Kolder, V. E. B., Gallagher, J. and Parsons, M. T. (1987) 'Court-ordered obstetrical interventions', *New England Journal of Medicine*, 316 (19): 1,192–6.

Koval, R. (1990) 'The commercialisation of reproductive technology IVF', in J. A. Scutt (ed.) *Baby Machine: reproductive technology and the commercialisation of motherhood*, London: The Merlin Press.

Koval, R. and Scutt, J. A. (1990) 'Genetic & reproductive engineering – all for the infertile?', in J. A. Scutt (ed.) *Baby Machine: reproductive technology and the commercialisation of motherhood*, London: The Merlin Press.

Kuhn, T. S. ([1962] 1996) *The Structure of Scientific Revolutions*, 3rd edn, Chicago, IL: University of Chicago Press.

Larbalestier, J. (ed.) (2006) *Daughters of Earth: feminist science fiction in the twentieth century*, Middletown, CT: Wesleyan University Press.

Latour, B. (1983) 'Give me a laboratory and I will raise the world', in K. Knorr-Cetina and M. Mulkay (eds) *Science Observed: perspectives in the social study of science*, London: Sage.

—— (1987) *Science in Action: how to follow scientists and engineers through society*, Cambridge, MA: Harvard University Press.

—— (1988) *The Pasteurization of France*, trans. A. Sheridan and J. Law, Cambridge, MA: Harvard University Press.

Latour, B. and Woolgar, W. ([1979] 1986) *Laboratory Life: the construction of scientific facts*, with intro. by J. Salk, with new postscript by authors, 2nd edn, Princeton: Princeton University Press.
Law, J. (2002) *Aircraft Stories: decentering the object in technoscience*, Durham, NC: Duke University Press.
—— (2004) *After Method: mess in social science research*, London: Routledge.
Law, J. and Hassard, J. (1999) *Actor Network Theory and After*, Boston, MA: Blackwell.
Lazarus, D. (2003) 'E-fight: Gulf War games get 2nd chance', *San Francisco Chronicle* (23 March); available at www.sfgate.com/cgi-bin/article.cgi?file=/chronicle/archive/2003/03/26/BU2781 (accessed 07/06/07).
Lefanu, S. (1988) *In the Chinks of the World Machine: feminism and science fiction*, London: The Women's Press.
Lette, K. (1993) *Foetal Attraction*, London: Picador.
Leavitt, J. W. (1986) *Brought to Bed: child-bearing in America, 1750–1950*, Oxford and New York: Oxford University Press.
Levine, G. (2000) 'Foreword', in G. Beer, *Darwin's Plots: evolutionary narrative in Darwin, George Eliot and nineteenth-century fiction*, 2nd edn, Cambridge: Cambridge University Press.
Lewis, S. ([1924] 1953) *Arrowsmith*, with an afterword by M. Schorer, New York: New American Library.
Lorber, J. W. (1988) 'In vitro fertilization and gender politics', in E. H. Hoffman Baruch, A. F. D'Adamo Jr. and J. Seager (eds) *Embryos, Ethics and Women's Rights: exploring the new reproductive technologies*, New York and London: Harrington Park Press.
Lupton, D. ([1994] 1995) *Medicine as Culture: illness, disease and the body in Western societies*, London: Sage.
Lykke, N. (2002) 'Feminist cultural studies of technoscience and other cyborg studies: a cartography', in R. Braidotti, J. Nieboer and S. Hirs (eds) *The Making of European Women's Studies*, IV (November): 133–43.
—— (forthcoming) 'Feminist cultural studies of technoscience: portrait of an implosion', in A. Smelik and N. Lykke (eds) *Bits of Life: feminist studies of media, biocultures and technoscience*, Seattle: University of Washington Press.
Lykke, N. and Braidotti, R. (1996) *Between Monsters, Goddesses and Cyborgs: feminist confrontations with science, medicine and cyberspace*, London: Zed Books.
McGuigan, J. (1992) *Cultural Populism*, London: Routledge.
MacKenzie, D. and Wajcman, J. (eds) (1999) *The Social Shaping of Technology*, Buckingham and Philadelphia: Open University Press.
McNeil, M. (1987) *Under the Banner of Science: Erasmus Darwin and his age*, Manchester: Manchester University Press.
—— (1988) 'Newton as national hero', in J. Fauvel, R. Flood, M. Shortland and R. Wilson (eds) *Let Newton Be! A new perspective on his life and works*, Oxford: Oxford University Press.
—— (1993) 'New reproductive technologies: dreams and broken promises', *Science as Culture*, 3 (4), no. 17: 483–506.
—— (1997) 'Clerical legacies and secular snares: patriarchal science and patriarchal science studies', *European Legacy*, I (5): 1,728–39.
—— (2001) 'Feminist reproductive politics: towards the millennium', in M. Conrad (ed.) *Active Engagements: a collection of lectures by holders of Nancy's Chair in Women's Studies, 1986–1998*, Halifax, NS: Mount Saint Vincent University.
—— (forthcoming) 'Roots and routes: the making of feminist cultural studies of technoscience', in A. Smelik and N. Lykke (eds) *Bits of Life: feminist studies of media, bioculture & technoscience*, Seattle: Washington University Press.

McNeil, M. and Castañeda, C. (eds) (2005) *Science as Culture*, 14 (2), special issue: Postcolonial Technoscience.
McNeil, M. and Franklin, S. (1991) 'Science and technology: questions for cultural studies and feminism', in S. Franklin, C. Lury and J. Stacey (eds) *Off-Centre: Feminism and Cultural Studies*, London: HarperCollins Academic.
—— (1993) 'Editorial: procreation stories', in *Science as Culture*, 3 (4), no. 17: 477–82.
Maddox, B. (2002) *Rosalind Franklin: the dark lady of DNA*, London: HarperCollins.
Marcus, G. (1995) *Technoscientific Imaginaries: conversations, profiles, and memoirs*, Chicago and London: University of Chicago Press.
Markus, G. (1987) 'Why is there no hermeneutics of natural sciences? Some preliminary theses', *Science in Context*, 1 (1): 5–51.
Martin, E. ([1987] 1992) *The Woman in the Body: a cultural analysis of reproduction*, 2nd edn, with new intro., Boston: Beacon Books.
—— (1990) 'The ideology of reproduction: the reproduction of ideology', in F. Ginsburg and A. L. Tsing (eds) *Uncertain Terms: negotiating gender in American culture*, Boston: Boston Press.
—— (1998) 'Anthropology and the cultural study of science', *Science, Technology, & Human Values*, 23, 1 (Winter): 24–44.
Maynard, M. (1994) 'Methods, practice and epistemology: the debate about feminism and research', in M. Maynard and J. Purvis (eds) *Researching Women's Lives from a Feminist Perspective*, London: Taylor and Francis.
Meikle, J. (2005) 'Women who delay babies until late 30s get health warning', *Guardian* (16 September); available at www.guardian.co.uk/medicine/story/0,11381,1571409,00.html.
Mellor, F. (2003) 'Between fact and fiction: demarcating science from non-science in popular physics books', *Social Studies of Science*, 33(4) (August): 509–38.
Menser, M. and Aronowitz, S. (1996) 'A manifesto on the cultural study of science and technology', in S. Aronowitz, B. R. Martinsons and M. Menser (eds) *Technoscience and Cyberculture*, New York: Routledge.
Merchant, C. (1980) *The Death of Nature: women, ecology and the scientific revolution*, San Francisco: Harper and Row.
Mialet, H. (1999) 'Do angels have bodies? Two stories about subjectivity in science: the cases of William X and Mister H', *Social Studies of Science* (August): 551–81.
Mies, M. (1986) 'Why do we need all this?', *Women's Studies International Forum*, 8 (6): 553–60.
Mirzoeff, N. (ed.) ([1998] 2002) *The Visual Culture Reader*, 2nd edn, London and New York: Routledge.
—— (1999) *An Introduction to Visual Culture*, London and New York: Routledge.
—— (2004) *Watching Babylon: the war in Iraq and global visual culture*, London and New York: Routledge.
Modell, J. (1989) 'Last chance babies: interpretations of parenthood in an in vitro fertilization programme', *Medical Anthropology Quarterly*, 3 (2): 124–38.
Mol, A. (2002) *The Body Multiple: ontology in medical practice*, Durham, NC: Duke University Press.
Monach, J. H. (1993) *Childless: No Choice: the experience of involuntary childlessness*, London: Routledge.
Moore, J. R. (1979) *The Post-Darwinian Controversies: a study of the Protestant struggle to come to terms with Darwin in Great Britain and America 1870–1900*, Cambridge: Cambridge University Press.
Moore, P. L. (1999) 'Selling reproduction', in E. A. Kaplan and S. Squier (eds) *Playing Dolly: technocultural formations, fantasies, & fictions of assisted reproduction*, London: Rutgers University Press.

Murdoch, A. (1990) 'Off the treadmill – leaving an IVF programme behind', in J. A. Scutt (ed.) *Baby Machine: reproductive technologies and the commercialization of motherhood*, London: The Merlin Press.
NASA (n.d. (a)) *John F. Kennedy Space Center: America's Spaceport*, NASA.
—— (n.d. (b)) *John F. Kennedy Space Center*, NASA.
—— (1992) *NASA Kennedy Space Center's Spaceport USA English Tourbook*, NASA.
Nature (1991) 'Women without men', *Nature*, 350 (14 March): 96.
Nelkin, D. (1996) 'Perspectives on the evolution of science studies', in S. Aronowitz, B. R. Martinsons and M. Menser (eds) *Technoscience and Cyberculture*, New York: Routledge.
Nelkin, D. and Lindee, S. M. (1995) *The DNA Mystique: the gene as a cultural icon*, New York: W. H. Freeman and Company.
Nelson, C., Treichler, P. A. and Grossberg, L. (1992) 'Introduction', in L. Grossberg, C. Nelson and P. Treichler (eds) *Cultural Studies*, New York and London: Routledge.
New Scientist (1991) 'Comment: Birth rights', *New Scientist*, 129 (16 March): 9.
Noble, D. (1992) *A World without Women: the Christian clerical culture of Western science*, New York: Alfred A. Knopf.
Norris, C. (1994) *Uncritical Theory: postmodernism, intellectuals and the Gulf War*, London: Lawrence and Wishart.
Novas, C. and Rose, N. (2000) 'Genetic risk and the birth of the somatic individual', *Economy and Society* (special issue on 'Configurations of Risk'), 29 (4): 484–513.
Nye, D. E. (1994) *American Technological Sublime*, Cambridge, MA, and London: MIT Press.
Oakley, A. (1976) 'Wisewoman and medicine man: changes in the management of childbirth', in J. Mitchell and A. Oakley (eds) *The Rights and Wrongs of Women*, Harmondsworth: Penguin.
—— (1984) *The Captured Womb: history of the medical care of pregnant women*, Oxford: Blackwell.
O'Brien, M. (1981) *The Politics of Reproduction*, London: Routledge & Kegan Paul.
Olby, R. (1974) *The Path to the Double Helix*, with a foreword by F. Crick. London: Macmillan.
Pearce, L. (2004) *The Rhetorics of Feminism: readings in contemporary cultural theory and the popular press*, London and New York: Routledge.
Penley, C. (1997) *NASA/Trek: popular science and sex in America*, London: Verso Books.
Penley, C. and Ross, A. (1991a) 'Introduction', in C. Penley and A. Ross (eds) *Technoculture*, Minneapolis: University of Minnesota Press.
Penley, C. and Ross, A. (eds) (1991b) *Technoculture*. Minneapolis: University of Minnesota Press.
Petchesky, R. P. (1987) 'Foetal images: the power of visual culture in the politics of reproduction', in M. Stanworth (ed.) *Reproductive Technologies*, Cambridge: Polity Press.
Petryna, A. (2002) *Biological Citizenship: science and the politics of health after Chernobyl*, Princeton: Princeton University Press.
Pfeffer, N. (1992) 'From private patients to privatization', in M. Stacey (ed.) *Changing Human Reproduction: social science perspectives*, London: Sage.
—— (1993) *The Stork and the Syringe: a political history of reproductive medicine*, Cambridge: Polity Press.
Phelan, P. (1997) *Mourning Sex: performing public memories*, London and New York: Routledge.
Plant, S. (1997) *Zeros and One: digital women and the new technology*, London: Fourth Estate.
Platt, S. (1991) 'Fertility control', *New Statesman and Society* (28 June): 11.
Plummer, K. (1995) *Telling Sexual Stories: power, change and social worlds*, London: Routledge.
—— (2003) *Intimate Citizenship: private decisions and public dialogues*, Seattle: University of Washington Press.

Pollitt, K. (1990) 'Fetal rights: a new assault on feminism', *The Nation*, 26 (March): 409–18.
Pope, A. ([1733] 1753) *An Essay on Man, in four epistles. With the notes of Mr Warburton*, London: J. & P. Knapton.
Powland, T. (1988) 'Reproductive technologies and the bottom line', in E. H. Hoffman Baruch, A. F. D'Adamo Jr. and J. Seager (eds) *Embryos, Ethics and Women's Rights: exploring the new reproductive technologies*, New York and London: Harrington Park Press.
Probyn, E. (1993) *Sexing the Self: gendered positions in cultural studies*, London and New York: Routledge.
—— (1996) *Outside Belongings*, London and New York: Routledge.
Rabinow, P. (1996) *Making PCR: a story of biotechnology*, Chicago: University of Chicago Press.
Rapp, R. (1999) *Testing Women, Testing the Fetus: the social impact of amniocentesis in America*, New York and London: Routledge.
Reinel, B. (1999) 'Reflections on cultural studies of technoscience', *European Journal of Cultural Studies*, 2 (2) (May): 163–89.
Rich, A. (1977) *Of Women Born: motherhood as an experience and institution*, London: Virago.
Richards, E. and Schuster, J. (1989a) 'The feminine method as myth and accounting resource: a challenge to gender studies and social studies of science', *Social Studies of Science*, 19: 697–720.
—— (1989b) 'So what's not a social category? Or You can't have it both ways (reply to Keller)', *Social Studies of Science*, 19: 725–29.
Riley, D. (1988) *Am I that Name? Feminism and the category 'woman' in history*, Basingstoke: Macmillan.
Roberts, R. (1993) *A New Species: gender and science fiction*, Urbana, IL: Illinois University Press.
Rose, H. (1994) *Love, Power and Knowledge: towards a feminist transformation of the sciences*, Cambridge: Polity.
—— (2002) 'In the shadow of men' (review of *Rosalind Franklin: the dark lady of DNA* by Brenda Maddox), *Guardian* (Review section) (15 June); available at www.books.guardian.co.uk/review/story/0,12084,737337,00.html (accessed 06/06/07).
Rose, H. and Rose, S. (1979) 'The radical science movement and its enemies', in R. Miliband and J. Saville (eds) *The Socialist Register*, London: Merlin Books.
Rose, N. (2001) 'The politics of life itself', *Theory, Culture & Society*, 18 (6) (December): 1–30.
Rosinsky, N. M. (1984) *Feminist Futures: contemporary women's speculative fiction*, Ann Arbor, MI: UMI Research Press.
Ross, A. (1989) 'Hacking away at the counter culture', in C. Penley and A. Ross (eds) *Technoculture*, Minneapolis: University of Minnesota Press.
—— (1991) *Strange Weather: culture, science and technology in the age of limits*, London: Verso.
—— (2000) *The Celebration Chronicles: life, liberty and the pursuit of property values in Disney's new town*, New York: Random House.
Rossiter, M. W. (1982) *Women Scientists in America: struggles and strategies to 1940*, Baltimore, MD: Johns Hopkins University Press.
—— (1995) *Women Scientists in America: before affirmative action 1940–1972*, Baltimore, MD, and London: Johns Hopkins University Press.
Rothman, B. K. (1986) *The Tentative Pregnancy: prenatal diagnosis and the future of motherhood*, New York: Viking.
Rouse, J. (1992) 'What are cultural studies of scientific knowledge?', *Configurations*, 1: 1–22.
Royal Commission on New Reproductive Technologies (1993) *Proceed with Care: final report of the Royal Commission on New Reproductive Technologies*, 2 vols, Ottawa: Minister of Government Services Canada.

Salinger, J. D. ([1951] 1958) *The Catcher in the Rye*, London: Penguin.
Samuels, S. U. (1995) *Fetal Rights, Women's Rights: gender equality in the workplace*, Madison: University of Wisconsin Press.
Sandelowski, M. (1993) *With Child in Mind: studies of the personal encounter with infertility*, Philadelphia: University of Pennsylvania Press.
Sawicki, J. (1991) *Disciplining Foucault: feminism, power, and the body*, London and New York: Routledge.
Sayre, A. (1975) *Rosalind Franklin & DNA*, New York: W. W. Norton & Co.
Scarry, E. (1985) *The Body in Pain: the making and the unmaking of the world*, Oxford: Oxford University Press.
Schiebinger, L. (1999) *Has Feminism Changed Science?*, Cambridge, MA: Harvard University Press.
Schneider, D. M. (1968) *American Kinship: a cultural account*, Englewood Cliffs, NJ: Prentice-Hall.
—— (1984) *A Critique of the Study of Kinship*, Ann Arbor, MI: University of Michigan Press.
Science and Technology Subgroup (1991) 'In the wake of the Alton Bill: science, technology and reproductive politics', in S. Franklin, C. Lury and J. Stacey (eds) *Off-Centre: feminism and cultural studies*, London: HarperCollins Academic.
Scott, J. (1992) 'Experience', in J. Butler and J. W. Scott (eds) *Feminists Theorize the Political*, New York and London: Routledge.
Shapin, S. (1990) 'Science and the public', in R. C. Olby, G. N. Cantor and J. R. R. Christie (eds) *Companion to the History of Modern Science*, London: Routledge.
—— (1992) 'Discipline and bounding: the history and sociology of science as seen through the externalism–internalism debate', *History of Science*, XX: 333–69.
Shapin, S. and Schaffer, S. (1985) *Leviathan and the Air-pump: Hobbes, Boyle and the experimental life*, including a translation of Thomas Hobbes, *Dialogus Physicus de Natura Aeris* by S. Schaffer, Princeton: Princeton University Press.
Shapiro, M. J. (1997) *Violent Cartographies: mapping cultures of war*, Minneapolis: Minnesota University Press.
Sherwin, S. (1995) 'Feminist ethics and in vitro fertilization', *Canadian Journal of Philosophy* (supplementary volume 13): 265–84.
Shohat, E. (1991) '"Laser for ladies": endo discourse and the inscription of science', *camera obscura*, 29: 57–89.
Shortland, M. and Yeo, R. (1996a) 'Introduction', in M. Shortland and R. Yeo (eds) *Telling Lives in Science: essays on scientific biography*, Cambridge: Cambridge University Press.
—— (eds) (1996b) *Telling Lives in Science: essays on scientific biography*, Cambridge: Cambridge University Press.
Showalter, E. (1997) *Hystories: hysterical epidemics and modern culture*, London: Picador.
Sismondi, S. (2004) *An Introduction to Science and Technology Studies*, Oxford: Blackwell.
Snitow, A. (1991) 'Feminism and motherhood: an American reading', *Feminist Review*, 40: 32–51.
Snow, C. P. ([1959] 1993) *The Two Cultures*, with intro. by S. Collini, Cambridge: Cambridge University Press.
Sourbut, E. (1997) 'Reproductive technologies and lesbian parents: an unnatural alliance?', in M. Maynard (ed.) *Science and the Construction of Women*, London: University College Press.
Spallone, P. (1989) *Beyond Conception: the new politics of reproduction*, London: Macmillan Education.
—— (1992) *Generation Games: genetic engineering and the future for our lives*, Philadelphia: Temple University Press.

—— (1996) 'The salutary tale of the pre-embryo', in Lykke, N. and Braidotti, R. *Between Monsters, Goddesses and Cyborgs: feminist confrontations with science, medicine and cyberspace*, London: Zed Books.
Spellman, E. V. (1988) *Inessential Woman: problems of exclusion in feminist thought*, Boston: Beacon Books.
Squier, S. (1999) 'Negotiating boundaries: from assisted reproduction to assisted replication', in E. A. Kaplan and S. Squier (eds) *Playing Dolly: technocultural formations, fantasies, & fictions of assisted reproduction*, New Brunswick, NJ, and London: Rutgers University Press.
Squires, J. (1996) 'Fabulous feminist futures and the lure of cyberculture', in J. Dovey (ed.) *Fractal Dreams: new media in social context*, London: Lawrence and Wishart.
Stabile, C. A. (1991) 'Shooting the mother: fetal photography and the politics of disappearance', *camera obscura*, 28: 179–205.
—— (1994) *Feminism and the Technological Fix*, Manchester: Manchester University Press.
Stacey, M. (1988) *The Sociology of Health and Healing: a textbook*, London: Unwin & Hyman.
—— (1992a) 'Conclusion', in M. Stacey (ed.) *Changing Human Reproduction: social science perspectives*, London: Sage.
—— (1992b) 'Introduction: what is the social science perspective', in M. Stacey (ed.) *Changing Human Reproduction: social science perspectives*, London: Sage.
—— (1992c) 'Social dimensions of assisted reproduction', in M. Stacey (ed.) *Changing Human Reproduction: social science perspectives*, London: Sage.
Stanley, L. (1992) *The Auto/biographical I: the theory and practice of feminist autobiography*, Manchester: Manchester University Press.
Stengers, I. (1997) *Power and Invention: situating science*, foreword by B. Latour, trans. P. Bain, Minneapolis: University of Minnesota Press.
Stent, G. S. (1980a) 'The author and publication of *The Double Helix*', in J. Watson *The Double Helix: a personal account of the discovery of the structure of DNA: text, commentary, reviews, original papers*, ed. Gunther S. Stent, New York and London: W. W. Norton & Co.
—— (1980b) 'A review of the reviews', in J. Watson *The Double Helix: a personal account of the discovery of the structure of DNA: text, commentary, reviews, original papers*, ed. Gunther S. Stent, New York and London: W. W. Norton & Co.
Strathern, M. (1992a) *After Nature: English kinship in the late twentieth century*, Cambridge: Cambridge University Press.
—— (1992b) *Reproducing the Future: essays on anthropology, kinship and the new technologies*, Manchester: Manchester University Press.
—— (1992c) 'The meaning of assisted kinship', in M. Stacey (ed.) *Changing Human Reproduction*, London: Sage.
Sturken, M. and Cartwright, L. (2001) *Practices of Looking: an introduction to visual culture*, Oxford: Oxford University Press.
Taylor, J. S. (1993) 'The public foetus and the family car: from abortion politics to a Volvo advertisement', *Science as Culture*, 3 (4), no. 17: 601–18.
Thomson, J. (1908) *The Complete Poetical Works of James Thomson*, edited with notes by J. L. Robertson, Oxford: Henry Froude.
Thompson, C. (2005) *Making Parents: the ontological choreography of reproductive technologies*, Cambridge, MA: MIT Press.
Thompson, E. P. ([1965] 1978) 'The peculiarities of the English', in *The Poverty of Theory and Other Essays*, London: Merlin Press.
Thornham, S. (2000) *Feminist Theory and Cultural Studies*, London: Arnold.

Throsby, K. (2004) *When IVF Fails: feminism, infertility and the negotiation of normality*, Basingstoke: Palgrave Macmillan.

Traweek, S. (1988) *Beamtimes and Lifetimes: the world of high energy physics*, Cambridge, MA: Harvard University Press.

Treichler, P. A. (1990) 'Feminism, medicine, and the meaning of childbirth', in M. Jacobus, E. F. Keller and S. Shuttleworth (eds) *Body/Politics: women and the discourses of science*, New York and London: Routledge.

—— (1991) 'How to have theory in an epidemic: the evolution of AIDS treatment activism', in C. Penley and A. Ross (eds) *Technoculture*, Minneapolis: University of Minnesota Press.

Treichler, P. A. and Cartwright, L. (1992a) 'Introduction', *camera obscura*, 28: 1–18.

—— (1992b) ' Introduction', *camera obscura*, 29: 5–17.

Tsing, A. L. (1990) 'Monster stories: women charged with perinatal endangerment', in F. Ginsburg and A. L. Tsing (eds) *Uncertain Terms: negotiating gender in American culture*, Boston: Beacon Press.

Turkle, S. (1996) *Life on the Screen: identity in the age of the internet*, London: Weidenfeld & Nicolson.

Turner, G. (2003) *British Cultural Studies*, 3rd edn, London: Sage.

Tyler, I. (2001) 'Skin-Tight: celebrity, pregnancy and subjectivity', in S. Ahmed and J. Stacey (eds) *Thinking through the Skin*, London: Routledge.

Urry, J. (1990) *The Tourist Gaze: leisure and travel in contemporary societies*, London: Sage.

van der Ploeg, I. (2001) *Prosthetic Bodies: the construction of the fetus and the couple in reproductive technologies*, Dordrecht: Kluwer Academic Publishing.

Van Dijck, J. (1998) *Imagination: popular images of genetics*, Basingstoke: Macmillan.

Van Wert, W. F. (1995–6) 'Disney World and posthistory', *Cultural Critique* (Winter): 187–214.

Vaughan, D. (1996) *The Challenger Launch Decision: risky technology, culture and deviance at NASA*, Chicago: University of Chicago Press.

Virilio, P. (1989) *War and Cinema: the logistics of perception*, trans. P. Camiller, London: Verso.

Virilio, P. and Lotringer, S. (1997) *Pure War*, revised edn, trans. M. Polizzotti, postscript translated by B. O'Keeffe, New York: Semiotext(e).

Voluntary Licensing Authority (1988) *Third Report of the Voluntary Licensing Authority*, London: Voluntary Licensing Authority.

Wajcman, J. (2000) ' Reflections on gender and technology studies: in what state is the art?', *Social Studies of Science*, 30 (3) (June): 447–64.

—— (2004) *Technofeminism*, Milton Keynes: Open University Press.

Waldby, C. and Mitchell, R. (2006) *Tissue Economies: blood, organs and cell lines*, Durham, NC, and London: Duke University Press.

Walkerdine, V. and Lucey, H. (1989) *Democracy in the Kitchen: regulating mothers and socialising daughters*, London: Virago.

Wall, A. (2000) 'Mothers, monsters and family values', in J. Marchessault and K. Sawchuk (eds) *Wild Science: Reading Feminism, Medicine and the Media*, London: Routledge.

Walsh, J. (1995a) 'Introduction: the legacy of the Gulf War', in J. Walsh (ed.) *The Gulf War Did Not Happen: politics, culture and warfare post-Vietnam*, Aldershot: Arena (Ashgate Publishing Ltd).

—— (1995b) *The Gulf War Did Not Happen: politics, culture and warfare post-Vietnam*, Aldershot: Arena (Ashgate Publishing Ltd).

Warnock, M. (1984) *A Question of Life* (reprint of *The Committee of Inquiry into Human Fertilisation and Embryology* 1984 Report), London: HMSO.

Wasko, J. (2001) *Understanding Disney: the manufacture of fantasy*, Cambridge: Polity.

Watkins, J. F. (1981) 'The laboratory life' (review of June Goodfield, *An Imagined World: a story of scientific discovery*), *Times Literary Supplement* (18 September): 1,058.

Watson, J. ([1968] 1969) *The Double Helix*, New York: Mentor.
—— (1980) *The Double Helix: a personal account of the discovery of the structure of DNA: text, commentary, reviews, original papers*, ed. Gunther S. Stent, New York and London: W. W. Norton & Co.
Weber, C. (2006) *Imagining America at War: morality, politics and film*, London: Routledge.
Wellings, K. (1985) 'Help or hype: an analysis of media coverage of the 1983 "pill scare"', in D. Leathar, G. B. Hastings, K. O'Reilly and J. K. Davies (eds) *Health Education and the Media II*, Oxford: Pergamon.
Wertenbaker, T. (1995) *The Break of Day*, London: Faber & Faber.
Whitbeck, C. (1988) 'Fetal imaging and fetal monitoring: finding the ethical issues', in E. H. Hoffman Baruch, A. F. D'Adamo Jr. and J. Seager (eds) *Embryos, Ethics and Women's Rights: exploring the new reproductive technologies*, New York and London: Harrington Park Press.
Wild, L. (2000–1) 'Trying: Leah Wild's IVF Diary' (June 2000–March 2001); available at www.guardian.co.uk/parents/story/0,3605,334535,00.html (17/05/07).
—— (2001) 'IVF diary', *Guardian* (March 14) (G2 section).
Williams, K. (2001) 'Designer babies or the birth of hope?', *Mail on Sunday: You Magazine* (8 May): 38–40.
Williams, R. (1973) *The Country and the City*, London: Chatto & Windus.
—— (1974) *Television: technology and cultural form*, London: Fontana.
—— ([1972] 1980) 'Ideas of nature', in *Problems in Materialism and Culture: selected essays*, London: Verso.
—— (1983) *Keywords: a vocabulary of culture and society*, revised and expanded edn, London: Fontana.
—— (1989) *Towards 2000*, London: Chatto & Windus.
Wilson, R. and Dissanayake, W. (eds) 1996 *Global/Local: cultural production and the transnational imaginary*, Durham, NC, and London: Duke University Press.
Winston, R. M. L. (1987) 'In vitro fertilization: practice, prospects and problems', in P. Byrne (ed.) *Medicine in Contemporary Society: King's College Studies 1986–7*, London: King Edward's Hospital Fund for London.
Winterson, J. (2001) 'Women have swallowed the lie that it is their right to have everything – including babies', *Guardian* (26 June): (Review section) 11.
Women's Studies Group (University of Birmingham Centre for Contemporary Cultural Studies) (eds) (1978) *Women Take Issue: aspects of women's subordination*, London: Hutchinson.
Woollett, A. (1996) 'Infertility: from "inside/out" to "outside/in"', *Feminism and Psychology*, 6 (1): 474–8.
Wright, P. (1985) *On Living in an Old Country: the national past in contemporary Britain*, London: Verso.
Wynne, B. (1995) 'Public understanding of science', in S. Jasanoff, G. E. Markle, J. C. Petersen and T. Pinch (eds) *Handbook of Science and Technology Studies*, London: Sage.
Yeo, R. (1988) 'Genius, method, and morality: images of Newton in Britain, 1760–1860', *Science in Context*, 2 (2) (Autumn): 257–84.
Yoxen, E. (1983) *The Gene Business: who should control biotechnology?*, London and Sydney: Pan Books.
—— (1985) 'Speaking out about competition: an essay on *The Double Helix* as popularisation', in T. Shinn and R. Whitley (eds) *Expository Science: forms and functions of popularisation*, Dordecht, Boston and Lancaster: D. Reidel.
Yuval-Davis, N. (1997) *Gender & Nation*, London, Thousand Oaks, CA, and New Delhi: Sage.
Zukin, S. (1991) *Landscapes of Power: from Detroit to Disney World*, Berkeley and Oxford: University of California Press.

Index

abortion 20
activism 17, 18, 20–21, 24, 146, 150
Actor Network Theory 148, 154n.4, 156n.21
Adorno, T. 141
Agamben, G. 161n.3
AIDS 20, 95
Akenside, Mark 29–30
Althusser, L. 19
Alton Bill 20
'Amanda's Story' 97–98
American Technological Sublime (Nye) 118
Amis, K. 47
Anderson, B. 29, 42, 117, 118, 127
Anderson, P. 33
anthropology 3, 12–14, 104
anti-heroes 49–51
Appadurai, A. 118–19
Aronowitz, S. 2–3
assisted reproductive technologies (ARTs) 74
Atwood, M. 146
autobiographies 51–52, 73, 110, 148

Bacon, Francis 33
Balsamo, A.: *camera obscura* 17; corporeality 129; cyborgs 143; disembodied technological gaze 122; female body 88; ideology 145; keeping watch 150
Barker, P. 127, 163n.11
Barthes, Roland 34
Bauman, Z. 161n.3
Beer, G. 14–16, 24
Berlant, L. 98, 100–102, 103
Billig, M. 36, 116–17, 128, 140–41
biographies 52–53, 65, 66–67, 138
biological citizenship 161n.3
biopolitics 83, 109, 161n.3
Birmingham University 18, 20

Blair, Tony 128
Blake, William 28–29, 30, 32–33, 37
Blood, D. 161n.9
Body/Politics (Jacobus) 16
Booth, A. 22
The Botanic Garden (Darwin) 28, 31
Boyle, Robert 65–66, 67, 137, 138
Bragg, M. 41
British cultural studies 18–21
Brito, Anna 53, 155n.9; *see also An Imagined World* (Goodfield)
Brooke, Henry 29
Brown, L. 89
Brown, M. 78
Bruno, G. 17–18
Bryld, M. 108, 109
Bryman, A. 140
Bush, George, Snr 119
Bush, George W. 130

camera obscura 17
Canadian Royal Commission on New Reproductive Technologies 84
Capital (Marx) 19
careful science 60–62
Carpenter, L. 71
cartographic approach 4–5
Cartwright, L. 17, 18
Centre for Contemporary Cultural Studies 18, 20
Challenger 164n.5
childlessness 157n.4, 158n.24; *see also* new reproductive technologies (NRTs)
cinematic representations 65
citizenship, new forms of 98–102
colonialism 12–13
combat, disappearance of 126
Comfort, N. 67
competitive scientists 51

Cooter, R. 7, 43
corporeality 120–24, 129, 149–50
The Country and the City (Williams) 19
Crick, Francis 46, 47, 48, 49
Crick, Odile 48
Crowe, C. 88
cultural anthropology 3, 12–14, 104
cultural studies 3, 5, 18–21, 141
Cultural Studies (Grossberg) 2
cultural studies of science and technology: doing 147–52; key dimensions 3; reflections on 41–43; versions of 2–4
cyberculture 2
cyberfeminism 142–43
cyberfictions 154n.6
Cyborg Babies (Davis-Floyd) 13
Cyborg Manifesto (Haraway) 22, 141
cyborg soldiers 127–28

Daily Mail 81, 107
Darwin, Charles 14–16, 33
Darwin, Erasmus 28–29, 31–33, 36–37
Darwinism 137
Darwin's Plots (Beer) 14–16
Davis-Floyd, R. 13
Dawson, G. 124–25
death counts 163n.9
Desmond, A. 15
diffraction 144
discovery narratives, spontaneous emergence and 1–2
Disney World 134, 140
domestic technology 89
donor insemination 75
The Double Helix (Watson): autbiographical form 51–52; competitive scientists 51; feminist anti-hero 49–51; feminist threat 57, 60; heteronormative restrictions 54; heteronormative science 48; heterosexual prowess 49; modest witness 66–67; narratives of discovery 60–61; overview 8, 44–45; rebuttal of 46, 61; rewriting scientific heroism 46–47; science as a man's world 49; scientist as ordinary guy 47; scientists and sexuality 56; secular, manly heroes 47–48; as sociological source 62–63
Doyal, L. 160n.46
dream narratives: older feminist dream narrative 85–88; overview 148; reproductive daydream narrative 89–90; true reproductive stories 79–84
Duden, B. 88, 116
Dumit, J. 13

Easlea, B. 136
Edge, D. 153n.5
Edwards, J. 13
Ehrenreich, B. 136
Einstein, Albert 34
elitism 141
English, D. 136
English national culture 33
Enloe, C. 125
environmental groups 21
EPCOT 134, 136, 145, 164n.7
epistemology 2–3, 19, 63, 149
Erickson, M. 8
An Essay on Man (Pope) 29
ethics of care 64, 156n.17
ethnography 13, 14, 19, 62–63, 64–65, 149
Evans, M. 148
Evans, N. 161n.9
exclusion 55
extraordinary lives 53–54

Fairbanks, C. 135
Faludi, S. 87, 92, 159n.35
Fara, P. 40–41
A Feeling for the Organism (Keller): careful science 62; extraordinary life 53–54; female exclusion 55; feminist threat 58–59; overview 45–46, 52; reviews 63–64, 67, 155n.9; scientists and sexuality 56–57
feminism: anthropology and 13; cultural studies 5–6; new reproductive technologies (NRTs) 85–88, 91; overview 149; post-feminism 67, 92; science studies and 64; second-wave feminism 85, 150; threat of 57–60; women watching 139–44
feminist anti-hero 49–51
feminist cultural studies of science and technology 3, 4–6
feminist science 144
feminist studies 4, 5
Feminist Theory and Cultural Studies (Thornham) 5
Fielden, K. 78
film studies 16–18
Firestone, S. 85, 91, 142–43
Flanagan, M. 22
Florida techno-tourism: *see* techno-tourism
Foucault, M. 20, 104, 158n.19, 161n.3
Fox, R. 121
Frankenstein (Shelley) 21

Frankfurt School 19, 141
Franklin, Rosalind: feminist anti-hero 49–51; science as a man's world 49; *see also Rosalind Franklin & DNA* (Sayre)
Franklin, S. 5, 6, 13, 74, 106
Frayling, C. 65
Freudian psychology 86
front line, disappearance of 126
FRONTLINE 115

gender: female exclusion 55; heteronormative restrictions 54; heteronormative science 48; heterosexual prowess 49; homosocial networks 55; science as a man's world 49
gender studies 16
genealogy: *see* roots and routes
Gilligan, C. 156n.17
Ginsburg, F. 13
Globe and Mail 145
Goodfield, June 62, 155n.6; *see also An Imagined World* (Goodfield)
Gray, A. 19
Gray, C. H. 113, 126–27
'Great Scientists' 41
Guardian 96–97
Guerlac, H. 27
The Gulf War (BBC and FRONTLINE) 115–16, 119–20, 129–30
Gulf War I: aftermath 120–24; coda 130–31; demise of soldier hero 124–28; overview 8–9, 113–15; spectacular nationalism 115–20
Gulf War II 130, 131
'Gulf War: Operation Desert Hammer' 124, 129–30, 163n.10
Gulf War Syndrome 124, 129
Gusterson, H. 123

Haraway, Donna: absent presences 6; agency 55; cyberfeminism 141; diffraction 144; feminist threat 60; modest witness 66; power relations 24; science as stories 71; science fiction studies 21–22; security clearances 136; technoscience studies 8, 9; women watching 137, 138, 139
Harding, S. 64, 142
Hartouni, V. 17
Harvey, P. 163n.3
heritage culture 37
heroes: demise of soldier hero 124–28; feminist anti-hero 49–51; new reproductive technologies (NRTs) 162n.14; rewriting scientific heroism 46–47; secular, manly heroes 47–48
heteronormative restrictions 54–55
heteronormative science 48
heterosexual prowess 49
Hewison, R. 37
Hird, M. 146
history of science 38–39
Hobbes, Thomas 65–66
homosocial networks 55
Hoskins, A. 117
Hotchkiss, R. 58
Hubbard, R. 73–74
Human Fertilisation and Embryo Authority 84, 92, 108, 159n.30
humanism 65
Humm, M. 80
Huxley, T.H. 137

ideology 75, 140–41, 143, 149, 150
Ignatieff, M. 115, 124
imaginary communities 29, 31, 42, 117, 148
An Imagined World (Goodfield): careful science 61–62; extraordinary life 53; feminist threat 57–58, 59–60; heteronormative restrictions 54–55; overview 45–46, 52; scientists and sexuality 56
in vitro fertilization (IVF) 75–76, 77, 107, 157n.6, 161n.1; *see also* new reproductive technologies (NRTs)
infertility 71, 92; *see also* new reproductive technologies (NRTs)
International Society for Literature and Science 16
internet 105–6
intimate citizenship 98–102, 148

Jacobus, M. 16
Jarhead 130
Jordanova, L. 14, 16–17, 42, 154n.5
Journal of Visual Culture 163n.4

Kaplan, A. E. 159n.32
Keller, Evelyn Fox: *Body/Politics* 16; careful science 62; contemporary studies of science 153n.5; elitism 141; feminist science 64, 144; science and feminist studies 4; as scientist 155n.6; *see also A Feeling for the Organism* (Keller)
Kellner, D. 117, 120
Kember, S. 143, 146

Kennedy Space Center 134
kill ratio 163n.9
kinship studies 13
Knight, Richard Payne 30–31

La Neuropatologia 17–18
laboratory lives 62–63
Languages of Nature (Jordanova) 14
Latour, B. 8, 62–63, 156n.19, 156n.21
'Leah Wild's Diary' 96–97, 109
Leavitt, J. W. 76
lesbianism, NRTs and 107–9, 161n.7
Let Newton Be! (Fauvel) 27
Leviathan and the Air-Pump (Shapin) 65–66, 138
Levine, G. 15
Lindee, S. 91
literary studies of science 14–16
Lock, M. 13
Locke, John 30
Lotringer, S. 126
Lucey, H. 86
Lucky Jim (Amis) 47
Lupton, D. 160n.45
Lury, C. 5, 6
Lykke, N. 4, 11–12, 16

McClintock, B. 155n.9, 155n.13; *see also A Feeling for the Organism* (Keller)
MacKenzie, D. 162n.2
Maddox, B. 67
Mail on Sunday 97–98
Making Parents (Thompson) 74
man's world, science as 49
Markus, G. 38
Martin, E. 13, 88, 144, 165n.15
Marxism 18–19
masculinity 49
media and communication studies 3
media representations 96–98; differentiation 106–9; proliferation 104–6; resilience 103–4; *see also* Gulf War I
media studies 19
menopause 108
Menser, M. 2
Merchant, C. 136
Mirzoeff, N. 113, 115, 118, 131
Mitchell, R. 77
modest witness 66–67, 138
Mol, A. 146, 165n.13
Moore, J. 15
motherhood 86–87
mythology 34–35

narratives of discovery 60–62
NASA 133, 164n.2
nationalism 115–20, 128–29
Native American Women's Health Education Resource Center 83
Nelkin, D. 3, 91
New England Journal of Medicine 87, 90
New Experiments Physico-mechanical (Boyle) 137
new reproductive technologies (NRTs): dream narrative 1 79–84; fictional representations 159n.34; older feminist dream narrative 85–88; overview 13, 71–72, 90–93, 94–95, 150; range of 75; reproductive daydream narrative 89–90; reproductive rights 161n.4; sibling donors 162n.11; terminology 73–76; ways of telling 76–79; *see also* tales of reproduction
New World Order 119–20, 128
Newton, Isaac: apple myth 34–35; in British culture 36–37; English national culture 33; in English popular culture 35–36; figure of scientist 38–39, 148–49; making of Newton 40–41, 43; modes of research 27–28; overview 148; poetry 28–33; scientific culture 39–40
Newton: the making of genius (Fara) 40–41
Nobel Prize 45, 46, 52, 156n.13
Noble, D. 136
Norrish, R. G. W. 57
Novas, C. 161n.3
Nye, D. 118, 136

Observer 122
Off-Centre: feminism and cultural studies (Franklin) 5
ontological choreography 77
Oprah Winfrey Show 105
Opticks (Newton) 29
ordinary guy, scientist as 47
other 14, 149

Parenthood (film) 87
Pasteur, Louis 156n.21
Penley, C. 17, 20, 21, 154n.3
Persian Gulf War 162n.1
Pfeffer, N. 78, 158n.16, 160n.47
PGD (pre-implantation genetic diagnosis) 97, 98, 106–9, 161n.1
Phelan, P. 120
Plant, S. 142–43
Platt, S. 91

Plummer, K. 98, 99–102, 103, 104–5, 106, 161n.2
political allegiances 5, 153n.6, 153n.7
Pope, Alexander 29
post-feminism 67, 92
postcolonial studies 153n.3
Potter, E. 137
power relations 24
pre-implantation genetic diagnosis (PGD) 75, 96–97, 107
pregnancy 76, 88, 89, 162n.11
Primate Visions (Haraway) 22, 60
Probyn, E. 73, 140
Proceed with Care (Royal Commission on New Reproductive Technologies) 95–96
public engagement 7, 151, 152
public understanding of science: biographies 65; emergence of 156n.22; importance of 7; poetry and 37; popular culture and 43; public engagement and 151, 152
Pumphrey, S. 7, 43
Purdy, J. 85
Pure War (Virilio) 124, 126

The Queen of America Goes to Washington City (Berlant) 100–102

Rabinow, P. 154n.9, 161n.3
Rapp, R. 13
Rashbrook, P. 159n.40
Reagan administration 100–102
Reload (Flanagan) 22
reproductive science: *see* new reproductive technologies (NRTs)
reproductive tourism 108
Richards, E. 64–65
Riley, D. 72
Roberts, E. 85
Rockefeller Foundation 56
roots and routes: British cultural studies 18–21; coda 22–24; cultural anthropology 12–14; overview 7–8, 11–12; science fiction studies 21–22; visual culture studies 16–18
Rosalind Franklin & DNA (Sayre): careful science 61; extraordinary life 53; female exclusion 55; feminist anti-hero 50; feminist threat 57, 59; heteronormative restrictions 54; overview 45, 52; scientists and sexuality 56
Rose, H. 154n.5
Rose, N. 161n.3

Ross, A. 20, 21, 154n.3
Rossiter, M. 136
Rouse, J. 153n.2
Royal Commission on New Reproductive Technologies 95–96

Saddam Hussein 131
Sandelowski, M. 88, 106
Saving Private Ryan 120–21, 124, 129–30
Sawicki, J. 87
Sayre, Anne 62, 155n.9; *see also Rosalind Franklin & DNA* (Sayre)
Scarry, E. 120, 121–23, 124, 127–28, 163n.9
Schaffer, S. 65–66, 136, 138–39
Schiebinger, L. 141
Schneider, D. 13
Schuster, J. 64–65
Schwarzkopf, General Norman 113, 116, 129
science: careful science 60–62; cinematic representations 65; competitive scientists 51; extraordinary lives 53–54; female exclusion 55; feminist threat 57–60; heteronormative restrictions 54–55; history of 38–39; as a man's world 49; public engagement 151–52; receptions and refractions 62–67; scientist as ordinary guy 48; scientists and sexuality 60; as stories 71; women watching 136–39
science fiction studies 21–22, 24, 154n.4
Science in Action (Latour) 63
science studies 4, 64
Scientific Revolution 138–39
Scott, J. 72
Scutt, J. A. 78
second-wave feminism 85, 150
security clearances 136
Sexing the Self (Probyn) 73
Sexual Stories (Plummer) 99
Sexual Visions (Jordanova) 16–17
sexuality: ARTs 108; heterosexual prowess 49; NRTs 107–9, 161n.7; scientists 55–57, 60
sexuality studies 16
Shalins, M. 13
Shapin, S. 65–66, 136, 138–39, 149, 153n.1
Shapiro, M. 121, 126
Shelley, M. 21
Shohat, E. 17
Shortland, M. 52
Showalter, E. 123–24

Shuttleworth, S. 16
Smart, Christopher 30
Snitow, A. 92
Snow, C. P. 15
social studies of science 62–64; history of 1
soldier heroes, demise of 124–28, 129
Spaceport USA 134, 135
Spallone, P. 91, 157n.1
Spielberg, Steven 120
Squires, J. 142–43
Stabile, C. 17, 79, 89, 125, 159n.37
Stacey, J. 5, 6
Stacey, M. 158n.23
Star Trek 21
Stengers, I. 146
Steptoe, P. 85
Strathern, M. 13, 74, 92, 104, 158n.17
structuralism 18–19
successor science 142
surgical strikes 121–22
surrogacy 75, 89, 105

tales of reproduction: differentiation 106–9; intimate citizenship 98–102; media representations 96–98; overview 9, 109–10; proliferation 104–6; resilience 103–4
techno-tourism: overview 132–34; reflections on 135–36; revisiting 144–45; women watching 136–39
techno-triumphalism 144
Technoscience and Cyberculture (Menser) 2
technoscience studies 9–10
terrorism 131
Thompson, C.: ARTs 74, 83; biomedical citizenship 161n.3; critical feminist perspectives 157n.9; heroes 162n.14; ontological choreography 77; reproductive rights 161n.4; sexuality 108
Thompson, E. 33, 34
Thomson, James 29–30
Thornham, S. 5
Three Kings 130
Throsby, K. 75, 76, 77, 84, 86, 89; IVF 107
Traweek, S. 65
Treichler, P. 17

Tsing, A. L. 159n.36
Turkle, S. 105

Van der Ploeg, I. 88
Van Dijck, J. 44, 51
Van Wert, W. F. 165n.9
Verran, H. 13
Vietnam War 121
Virgin Birth controversy 81–82, 107
Virilio, P. 114, 120, 124, 126, 129–30
A Vision of the Last Judgement (Blake) 30
visual culture studies 16–18, 163n.4
Voluntary Licensing Authority 76, 84

Wajcman, J. 162n.2
Waldby, C. 77
Walkerdine, V. 86
Walsh, J. 117, 122
war games 124, 127–28
warfare: *see* Gulf War I; World War II
Warnock Committee 75, 78
Watson, Elizabeth 48
Watson, James 44; *see also The Double Helix* (Watson)
Weber, C. 120
Wilberforce, Bishop S. 137
Wild, Leah 96–97, 161n.1
Wilkins, M. 46, 49, 155n.4
Williams, R. 19
Winston, R. 160n.41
Winterson, J. 94, 110
women in Gulf War I 125–26
Women Take Issue (Women's Studies Group) 5
women watching 136–39, 150; feminist views of technoscience 139–44
women's experience 72
Women's Studies Group 5
Woolgar, S. 62–63
World Trade Center attacks 131
World War I 127
World War II 120
A World without Women (Noble) 136
Wright, J. 137

Yeo, R. 39–40, 41, 52
Yoxen, E. 51
Yuval-Davies, N. 125, 127

Lightning Source UK Ltd.
Milton Keynes UK
23 April 2010

153193UK00004BB/9/P